ECOLOGY DOCUMENTARIES

This companion piece to Susan Hayward's *Film Ecology* focuses on ecology documentaries produced in the first 20 years of the new millennium (2000–19).

Using Kate Raworth's regenerative economic theoretical model as set out in *Doughnut Economics*, this book examines 57 films emanating from Europe and the 4 areas of concern they raise about energy production, pollution and waste management, agribusiness, disrupted ecosystems and the migratory flow. These ecology documentaries make explicit the damage done to our planet thanks to growth capitalism and neoliberal globalisation. But they also provide the evidence that solutions to this planetary abuse exist. The book demonstrates how these documentaries reveal the process of humankind's planetary plundering and explores the structuring of the eco-doc as a new generic type in the domain of documentary practice. Using Raworth's model allows us to measure the tentacular extent of the planetary harm growth economics induces and, too, by way of contrast, perceive how regenerative economics can work to redress this harm, heal the Earth and make it a safe place for humanity.

This book is ideal for film studies scholars and students, including those teaching or studying film practice, documentary film, European cinema and environmental studies, as well as economists interested in regenerative economic models. It also has general appeal to all who are concerned about some of the major causes of planetary degradation and its impact on humanity and Earth.

Susan Hayward is Emerita Professor of Cinema Studies at Exeter University, UK. She is the author of several books on French cinema and *Cinema Studies: The Key Concepts* (now in its fifth edition).

ECOLOGY DOCUMENTARIES

Their Function and Value Seen Through the Lens of Doughnut Economics

Susan Hayward

Routledge
Taylor & Francis Group
LONDON AND NEW YORK

First published 2022
by Routledge
2 Park Square, Milton Park, Abingdon, Oxon OX14 4RN

and by Routledge
605 Third Avenue, New York, NY 10158

Routledge is an imprint of the Taylor & Francis Group, an informa business

British Library Cataloguing-in-Publication Data
A catalogue record for this book is available from the British Library

Library of Congress Cataloging-in-Publication Data
A catalog record has been requested for this book

ISBN: 978-0-367-62192-6 (hbk)
ISBN: 978-1-032-13162-7 (pbk)
ISBN: 978-1-003-13024-6 (ebk)

DOI: 10.4324/9781003130246

Typeset in Bembo
by Deanta Global Publishing Services, Chennai, India

This book is in homage to the courage of those who made these films and in honour of those who seek to heal the Earth. It is also in homage to the many of the 'unheard' – most often the ones who first and foremost suffer the consequences of our anthropogenic global warming.

CONTENTS

Acknowledgements viii
List of figures ix
List of tables xi

Introduction 1

1 ENERGY – crude awakenings: Extractive energy sources versus regenerative energy practices 14

2 Waste … means … money … and suffocating the earth – the human practice of pollution 59

3 Bad-husbandry versus Earth-as-provender, a tale of two cultures: Agribusiness' monocultural-monopolies and Agroecology's pluriculturalism 84

4 The effects of disrupted ecosystems, the migratory flow and the unsafe, inhumane space of the unheard 119

Conclusion 168

Bibliography 173
Appendix A: List of 57 European ecology-documentaries 2000–19 175
Index 182

ACKNOWLEDGEMENTS

My thanks go to Gábor Gergely who set me on the path of ecology documentaries when he asked me to contribute to his book *Companion to European Cinema*. His thoughtful feedback on my chapter was what drove me to develop it into a full-length book study. I am very grateful indeed to Chrissie Morris for her eagle-eyed editorial expertise and constructive comments which greatly improved the structure of Chapters 2 and 4. My thanks to Regenia Gagnier for the often-enlightening conversations on our country walks during the prolonged COVID-era. I am equally very appreciative of my ecology discussions with Caroline Ives. Thanks also to Christine Walters and MeiLi for our invaluable garden picnics. Finally, I am, as before (with *Film Ecology*), enormously indebted to Kate Raworth's groundbreaking study, *Doughnut Economics: Seven Ways to Think Like a 21st-Century Economist* – her theoretical model is the guiding light of this study.

FIGURES

I.1 Charting the European eco-docs 2000–19 3
I.2 Raworth's basic Doughnut diagram 8
I.3 Raworth's detailed Doughnut diagram 9
I.4 Mapping the 57 eco-docs onto the Doughnut, a visualisation of planetary degradation and critical human deprivation 11
1.1 Mapping the seven 'cheap energies' eco-docs onto the Doughnut 37
1.2 Raworth's butterfly diagram adapted from Ellen MacArthur's model in *Towards a Circular Economy* 40
1.3 Triangulation of energy autonomy 41
1.4 juwi Group headquarters, the world's most energy-efficient building 43
1.5 Triangulation of the processes of empowerment and cultural contexts within *Solar Mamas* 47
1.6 Mapping *The 4th Revolution* (4thR) and *Solar Mamas* (SM) onto the Doughnut 49
2.1 Mapping our wasteful ways: kilograms of solid waste generated per person per year 61
2.2 Mapping the pollution eco-docs' performative documentary practice and their visualising "the beyond control" – plotting the transgressions of the social and planetary boundaries by waste-management networks and the impact of their practices on local inhabitants 72
2.3 Raworth's diagram of the degenerative linear economy 73

2.4 Raworth's diagram of the Ellen MacArthur Foundation's butterfly model for a circular economy 73
3.1 UN Food and Agricultural data – global land use 86
3.2 Mapping the agribusiness eco-docs onto the Doughnut 98
3.3 The regenerative and democratic value of agroecology 103
3.4 Agricultural autonomy as a source of equality and democracy – as promoted by agroecology 111
4.1 The UNHCR figures of forcibly displaced people at a glance 2019 122
4.2 Israeli checkpoints in Gaza Strip. Zeitoun the location of *Samouni Road* is due south of Gaza City 134
4.3 Occupied Palestinian Territory. Area C in light blue is the part of the West Bank under full Israeli control 135
4.4 A map of the West Bank showing Israeli border crossing points, internal checkpoints and security control areas; also shown is No Man's Land 136
4.5 Map showing Bil'in and Ni'lin. The outlines in green indicate the 'No Man's Land Area'. In red: the separation barrier. In pale blue: the existing Modi'in Illit (Israeli settlements) bloc's jurisdictional area, built on formerly owned Palestinian land. In yellow: the areas included in the Israeli master-plan for further settlements that lie outside the jurisdictional area 138
4.6 Amal 143
4.7 The European Union's 'wall-building' 150
4.8 Mapping the nine eco-docs on disrupted ecosystems onto the Doughnut 161

TABLES

I.1 Figures for single-country productions of eco-docs 4
1.1 Table of energy eco-docs 15
2.1 Table of eco-docs about pollution from domestic, medical and industrial refuse 60
3.1 Table of Agribusiness and Agroecology eco-docs 85
4.1 Table of eco-docs about effects of disrupted ecosystems and global warming 120
4.2 The five episodes of *5 Broken Cameras* 140

INTRODUCTION

The objective behind this book is to offer a companion (or follow-up) piece to my *Film Ecology* book (Hayward, 2020). This study will take forward the theoretical ideas developed in *Film Ecology: Defending the Biosphere – Doughnut Economics and Film Theory and Practice*, which are based on Kate Raworth's *Doughnut Economics* (2017), and focus this time on ecology documentaries produced over the period 2000–19 – a period which has seen a huge increase in documentaries world-wide about all aspects of climate change.

This rise in ecology documentary films (eco-docs) has been matched by an increased interest in teaching and writing about them – primarily seeing these products as resistances to the prevailing winds of climate denialism. Most of these studies explore the films' role in the 'socio-political struggle toward greater ecological responsibility' as Adam O'Brien puts it in *Film and the Natural Environment: Elements and Atmosphere* (2018: 9). The pitfalls inherent in this filmic role of the eco-doc, however, are not to be underestimated, as the editors of *Ecocinema Theory and Practice* (Rust et al., 2013) point out, arguing that documentaries about the planet – particularly those made by Hollywood, or for television – tend, still, to adopt an anthropocentric approach. While this is certainly one tendency, Helen Hughes, in her interesting study, *Green Documentary: Environmental Documentary and the Twenty-First Century* (2014), salutes the eco-doc's potential to act as a communicative response to environmental consciousness. Remarkably, she distinguishes four modalities of awareness through which the eco-doc delivers its message (contemplative, ironic, argumentative, material) – thereby implicitly seeing the eco-doc as firmly linked to deep-consciousness which includes slowing down information and allowing the layers of the issues to emerge, and, at the same time, acknowledging the paradox inherent in the eco-doc's own practice of filmmaking. For, just as it draws attention to environmental damage, so too it has a toxic materiality of its own (12). This latter theme is dealt with in great detail in Nadia Bozak's *The Cinematic Footprint: Lights, Camera, Natural Resources* (2011) and Mette Hjort's *Film and Risk* (2012). Lastly, in terms of books I have found helpful, John A. Duvall (2017) has produced an invaluably detailed descriptive inventory of documentary activism in the new millennia with useful information on distribution practices in *The Environmental Documentary:*

DOI: 10.4324/9781003130246-101

Cinema Activism in the 21st Century. In terms of journals, a very important source is Intellect Press' *Studies in Documentary Film*.

There has already been a great deal written about documentaries as a genre, but this study is adopting a different approach, more readily identifiable as embedded in social-science-economics. The intention here is to investigate the way in which eco-docs have responded to the planetary harm the human species has and is continuing to cause through a dysfunctional degenerative economic system – that is, growth economics, known alternatively as the Capitolocene or the Anthropocene. The purpose of this study, then, is to explore the value of a particular theoretical, economic model – Doughnut Economics – and to demonstrate how it is one that is in tune with the function and purpose of eco-docs. This approach is one that allows us to look at these two distinct economic models simultaneously in terms of their impact on the planet's environment, namely, the effects of the Anthropocene model (growth capitalism) and the potential of the regenerative model (Doughnut Economics).

I.1 Research corpus

When Anita Angelone (2011: 155) speaks of the 'performative exchange of documentary', we have in those few words the thrust of the roughly 163 *commercially* released eco-docs (or so) produced since the turn of this century. Fairly evenly distributed, in numerical terms, over the two decades, showing that the urgency of these testimonies has not diminished, these eco-docs are exhibited over a multiplicity of platforms – including cinema theatres, film festivals and pop-ups, television channels and streaming platforms on-line (most noticeably through Netflix, but also accessible on YouTube). Whilst the US dominates numerically with 42 percent of the output of eco-docs (69), significantly, 35 percent of these eco-docs (57) are European in terms of directors and producers, with a further 34 eco-docs emanating from Australia (7), Canada (17), China (1), Columbia (1), India (1), Indonesia (1) and Japan (3), and a further 4 being co-productions between non-European countries.

The choice with this book is to focus on the European eco-docs. This is not to gainsay the American output (nor indeed those of the other non-European countries); however, the US documentaries lean heavily towards coverage of environmental issues close to hand within their national borders (at 60 percent), whereas almost every European eco-doc deals with a range of issues that are global and encompasses all continents of the world. So my decision to focus on European eco-docs lies in the extent of their reach and in their ability to present the issues within a broad political, economic and historic framework. For example, it is clear that many of the environmental problems relating to nourishing the world's population (particularly in the African states, Latin American states and the Indian subcontinent) will find their earliest roots in the agricultural practices extant under colonialisation (large-scale plantations being the prime examples) instituted by European empire-building nations and the consequential effects under post-colonialism (see Hayward, 2020: 78–80).

In terms of European eco-docs, let's first take note of what has been produced. I have managed to trace 57 European eco-docs (all listed in Figure I.1).

1. Agribusiness/Agri-industry: bio-diversity loss and land or space conversion/pesticides-concentrated food production/Agroecology/solutions

Agribusiness: Bio-diversity loss, land or space conversion, agrochemical products, monocultures and concentrated food production	Agroecology/solutions
1. *Darwin's Nightmare* ***$m** Austrian/Fr/Bel 2. *Earth* UK/Ger/US **‡$m** 3. *Les Enfants nous accuseront/Our Children Will Accuse Us* Fr 4. *Humain/Human* Fr * 5. *Idiot Cycle* Fr/Can ‡ 6. *McLibel* UK * 7. *Le Monde selon Monsanto/ The World According to Monsanto* Fr ‡* 8. *More than Honey* ‡ **$m** Switzerland 9. *Round-up face à ses juges/Round-up Facing Its Judges* Fr 10. *Le Sel de la terre/The Salt of the Earth* Fr * **$m** 11. *Vanishing of the Bees* UK ‡	12. *Anaïs s'en va-t-en guerre/Anaïs Goes to War* Fr ‡ 13. *Après demain/After tomorrow* Fr 14. *Au nom de la terre/For the Sake of the Earth* Fr ‡ 15. *Cultures en transition/Voices of Transition* Fr/Ger ‡ 16. *Demain* Fr * **$m** 17. *Forests of the 21st Century UK* 18. *Forgotten District* Fr/UK/Ger 19. *Just Do It* UK 20. *Solutions locales pour un désordre global/Think Global, Act Local* Fr **$** 21. *La Vie moderne/Modern Life* Fr * **$m**

2. Disrupted ecosystems and the migratory flow: climate change/war, internally displaced persons/immigrants/refugees/migrants

Climate change	War/internally displaced persons	Refugees/immigrants/migrants
1. *The Age of Stupid* UK ‡ **$** 2. *La Glace et le ciel/Ice and the Sky* Fr *	3. *Five Broken Cameras* Fr/Israel/Netherlands/UK/Finland/ Switzerland/US/Can/Rep of Korea ‡***$** 4. *For Sama* UK ‡* **$m** 5. *Of Fathers and Sons* Ger/Syria * 6. *Samouni Road* It/Fr * 7. *White Helmets* UK *	8. *Broken Land* Switzerland ‡**$** 9. *Come un uomo sulla terra/Like a Man on Earth* It * 10. *Eldorado* Switzerland ‡* 11. *Fuocoammare/Fire at Sea* It * **$m** 12. *Human Flow* Ger ‡***$m** 13. *Il sangue verde/Green Blood* It * 14. *Sea Sorrow* UK *

3. Pollution: from domestic, medical and industrial refuse

Water	Air	Land
1. *Before Vanishing* Fr/Syria	2. *La bambina deve prendere aria/Baby Needs Some Fresh Air* It 3. *Clouds of Smoke* Albania/UK 4. *No Pressure* UK	5. *Biùtiful cauntri/Beautiful Country* It * **$** 6. *Una montagna di balle/Mountain of Lies* It 7. *Polluting Paradise* Ger * 8. *Trashed* UK * 9. *Waste Land* UK/Brazil * **$**

4. Energy: nuclear/oil/minerals/dams vs alternatives

Nuclear	Oil/minerals/hydropower	Alternatives
1. *The Day the Sun Fell* Switzerland ‡ 2. *How to Change the World* UK/Can ‡* 3. *Into Eternity* Den ‡ 4. *Journey to the Safest Place on Earth* Switzerland ‡	5. *A Crude Awakening: The Oil Crash* Switzerland ‡ 6. *Dreamland* Iceland **$** 7. *Home* Fr * **$** 8. *Stealing Africa* Den ‡ 9. *Undamaged* Slovenia 10. *Virunga* UK/Congo ‡*	11. *The 4th Revolution* Ger ‡ **$** 12. *Les Glaneurs et la glaneuse/The Gleaners and I* Fr ‡* **$** 13. *Solar Mamas* Den/UK/US ‡*

FIGURE I.1 Charting the European eco-docs 2000–19 (an asterisk denotes a major award, special selection or nomination; ‡ indicates minor awards and festivals; the $m sign = box-office returns over $1 million; single $ sign = box office over $100,000).

Whilst there may be some omissions, the list is strongly representative of what is currently out there and available to see.

In terms of output, 43 eco-docs are single-country productions; the figures break down as seen in Table I.1.

TABLE I.1 Figures for single-country productions of eco-docs

France	*UK*	*Switzerland*	*Italy*	*Germany*	*Denmark*	*Iceland*	*Slovenia*
14	10	6	6	3	2	1	1

The remaining 14 are co-productions, 5 co-produced by European countries, 9 co-produced between European countries and non-European countries (Syria, Congo, Canada, Brazil, South Korea and the USA).

Of these 57 eco-docs, 43 were in film festivals (major and minor) and won awards (see details in Appendix 1). Out of these numbers, 28 received major awards at major international film festivals – a significant achievement attesting once again to the pertinence of their subject matter. Of these 28 films, 9 were nominated for an Oscar (*White Helmets*, eventually a winner in 2017); five for the César award with four winners (*Darwin's Nightmare*, 2004; *Home*, 2009; *Le Sel de la terre/The Salt of the Earth*, 2015; *Demain*, 2015); two Academy of Italian Cinema Awards (*Come un uomo sulla terra/Like a Man on Earth*, 2008; *Samouni Road*, 2018); one Italian Golden Globe winner (*Biùtiful cauntri/Beautiful Country*, 2007); one BAFTA (*For Sama*, 2019); one Cannes Festival Special Prize (*Le Sel de la terre/The Salt of the Earth*, 2014); one Cannes best documentary (*Samouni Road*, 2018); one Cannes Trophée des sciences du danger (*Le Monde selon Monsanto/The World According to Monsanto*, 2008); one Golden Lion (*Human Flow*, 2017); one Golden Bear award (*Fuocoammare/Fire at* Sea, 2017), also at Berlin Film Festival the Amnesty International Film Prize for *Eldorado* (2018); and Sundance Film Festival World Documentary Grand Jury Prizes for *How to Change the World* (2015) and *Of Fathers and Sons* (2017). In terms of audiences, a third of these 57 eco-docs obtained good to outstanding box-office returns. Ten films exceeded the $1 million box office mark – four attained a range from $3 million to $11 million, and one in particular garnered $109 million (*Earth*, 2007). Of the rest, a further nine eco-docs achieved the hundred-thousand-dollar mark or more (for details on awards and revenue, see Figure I.1).

Once we come to defining the categories into which these eco-docs fall, there are four dominant areas within which several related themes emerge. Here is the list in numerical order (in parenthesis is the percentage of those receiving major international awards or nominations):

1. Agribusiness and Agroecology: 21 (33%)
2. Disrupted ecosystems and migratory flow: 14 (93%)
3. Pollution: 9 (33%)
4. Energy: 13 (38%)

You will note that the eco-docs focusing on disrupted ecosystems and the migratory flow are the most decorated, which surely says something about the perception of the refugee crisis as a profoundly important ecological issue (four out of the nine Oscar-nominated eco-docs are also in this category).

And here, charted in Figure I.1, are the four dominant categories broken down into greater detail (for further details see Appendix 1).

If we take a brief survey of this figure, we can see that France's main focus is on the devastating effects of giant agribusiness practices and the potential for answers to this in terms of regenerating the Earth through agroecology. Meanwhile, Italy's preoccupations are with two major issues: the reality and impact of immigrants arriving on their shores on the one hand and, on the other, pollution and corrupt waste management. The UK for its part presents its concerns in the first three categories (agribusiness, the effects of disrupted ecosystems and global warming on the planet and peoples, and pollution). Switzerland's focus, as with Germany, is on the plight of immigrants and energy issues. Finally, Denmark, Iceland and Slovenia look to the problems of current energy practices.

I.2 A few comments on the documentary practices at work in this corpus

Generally speaking, the duration of documentary-filming for our corpus of eco-docs – usually two to four years – is one that, obviously, allows for stories to unfold and to modify in synch with events; events that might, incidentally, change the original orientation of the film (see *The Day the Sun Fell* and *Virunga*, for example, discussed in Chapter 1). This lack of compression, in favour of slow-filming, makes the eco-doc an invaluable tool insofar as it can make visible what T. J. Demos in his book *Against the Anthropocene* calls the 'slow violence' – the slow-moving damage – done to the planet (for more on this, see Hayward, 2020: 17). But this durational element also allows witnesses to find their own means of expression, whether it is to speak to the harm done to their living environment or to themselves – an affording of space and time for their courage to manifest itself. Indeed, in virtually all the eco-docs (as we shall see) those people approached or interviewed by the filmmakers are more than witnesses; they become contributors, participants and collaborators. As an effect of slow-filming, of duration, their various forms of living find space.[1] Nor should it be overlooked that, in many instances, with these eco-docs, a great deal of courage is involved in the filmmakers' process of exposing this planetary devastation. In several films, we witness them going undercover, or using hidden cameras to record illegal undertakings or acts of police or military violence, or again facing dangerous, life-threatening conditions whilst trying to record the evidence.

A second point worth considering – given the economic approach of this study – is the finance involved, and in particular, who finances. Again, I shall only make some broad statements and illustrate with a few examples. Generally speaking, documentaries are cheaper to produce than narrative films. As a rule of thumb, a documentary is presently costed at $1000 per finished minute. Thus, budgets for most of the eco-docs listed in Figure I.1 could be estimated to range between \$600,000 and \$1 million. Of course a number were made on tight budgets – possibly as low as \$20,000–40,000, or lower. For example, *Undamaged*, a 52-minute documentary about river preservation in the Balkans, came in at just over \$5000.[2] For that reason, truly expensive eco-docs tend to stand out and, indeed, raise eyebrows as to both the practice and the financing. Just three of the eco-docs, *Earth* (Alastair Fothergill and Mark Linfield, 2007), *Home* and *Human* (2009 and 2016 respectively, both by Yann Arthus-Bertrand), fall into this category. The massive \$47 million-budget documentary *Earth* highlights the impact of climate change on species survival (with special focus on the polar bear, humpback whale and African bush-elephant). It box-officed \$109 million, justifying, one could say, the funding (from BBC Natural History Unit/UK, Greenlight Media/Germany, with Discovery Channel/USA making some financial contribution); and it is the highest recorded return on a nature documentary. The major sourcing of finance for this eco-doc comes from the UK and German producers whose environmental credentials are well-established. Discovery Channel is a subscription TV company specialising in reality-based series. There is little to quibble with here, it seems to me (and the film is exquisitely made). The Yann Arthus-Bertrand eco-docs did, however, raise critical comments, particularly *Home*, with regard both to the technology used and the financing (issues I address in Chapter 1 in my discussion of this film). The French press were especially eloquent in labelling Bertrand an *éco-tartuffe* (an eco-hypocrite) and a *hélicologiste* because of the helicopter equipment he used and the multinationals he drew upon to help fund his projects (in the case of *Home*, Kering and the Qatar Foundation; and with *Human*, the Bettancourt Foundation and his own GreenPlanet Foundation which counts amongst its donators EDF, Suez and Coca-Cola, not really known for their ecological practice, nor observance of human rights).[3] To his critics' way of thinking he was exploiting to his own ends the very economic model that ecologists fight against.

But these examples are exceptions. On the whole the financing is pulled together by a diverse set of transparent means. Some filmmakers run their own environmentally green production companies – such as Why Poverty in Denmark, Greenlight in Germany, Manaia Productions in France and Grain Media, Violet Films and Chouette Films in the UK. Others associate with producers who operate transnationally to obtain funding from public broadcasters (e.g., Arte, WDR and the Canadian NFB as with *Le Monde selon Monsanto*). Alternatively, filmmakers themselves become producers and source revenue from individuals or crowd-fund. Other elements that help are European networks of regional funds such as Cine-Regio and the initiative EcoProd with their green-filmmaking project.

One last comment: a particularly striking attribute of eco-docs in general and one borne out by the eco-docs in this corpus is, as Helen Hughes explains, not just their 'transnational capacity for environmental concern, a cosmopolitan consciousness developing alongside globalisation', but also that their work constitutes a series of 'experiments in documentary production, connecting individual responses to the larger themes' (2014: 137–8). She also cites Agnès Varda who saw her making of

Les Glaneurs et la glaneuse as 'a way of living, not just a product' (137), which makes me think of the anthropologist Kathleen Millar's concept 'forms of living' (2018: 9), whereby our eco-doc filmmakers are ecosystems that diverge from 'state-regulated officially recognised institutionally organised forms of work in capitalist societies' (13). That is to say, they do not function within the wage-labour construct of capitalism, any more than do their films – the amount of footage taken (often over 400 hours), the duration of the shooting (several years), the risks taken, the valuing of the voices of the unheard, all these are just some of the aspects of their forms of living. We shall come across many instances of this during the next four chapters of this book. But now onto the last section of this Introduction: Kate Raworth's *Doughnut Economics* model.

I.3 Introduction to the theoretical model: Doughnut Economics

In *Doughnut Economics*, Kate Raworth begins by explaining how our current economic model, which stretches back some 200 years, is an outcome of the Industrial Revolution and the concomitant belief in a growth model that can exist forever as if the planet had no finite resources (8–9). However, as she points out, the inevitability of the growth economy model is an ecological disaster, for the planet will run out of resources (31). Yet economists and politicians, particularly in Western nations, remain fixated – and have been even more so, after World War Two – 'on GDP, or national output … a fixation that has been used to justify extreme inequalities of income and wealth coupled with unprecedented destruction of the living world' (25). Since the 1980s, 'when Margaret Thatcher and Ronald Reagan teamed up to bring the neoliberal script to the international stage' (67), the economy has speeded up thanks to the explosion of free-market capitalism which has meant the liberalisation of financial markets and deregulation practices such as free trade, privatisation, and a diminution in government spending (often leading to austerity measures affecting the poor). This neoliberal economic model, which the West brought to the world under the guise that the 'market is the road to freedom' (70), has become the ideological driving force behind contemporary globalisation with its 'rapid expansion of cross-border flows in the last 20 years' which, while they 'have the potential to deliver benefits', also 'carry risks' (as Raworth explains, 89). Indeed, the fast economic growth of the twenty-first century has profited corporate interests and strengthened the power of industrialised nations. Clearly, the unrestricted free-trade benefits ultimately profit those nations with the financial clout. Thus, the economic gap between rich and poor countries continues to grow. However, it is also clear that 'the cumulative effects of economic growth and modernisation in industry' (which include energy-resources exploitation and the so-called Green Revolution of GMOs and Glyphosate) have deeply impacted upon the environment, so much so that global sustainability is now greatly at risk.

To Raworth's mind, it is evident that a linear economy (of growth) cannot last. So what must replace it is a circular, regenerative model (34). To this effect,

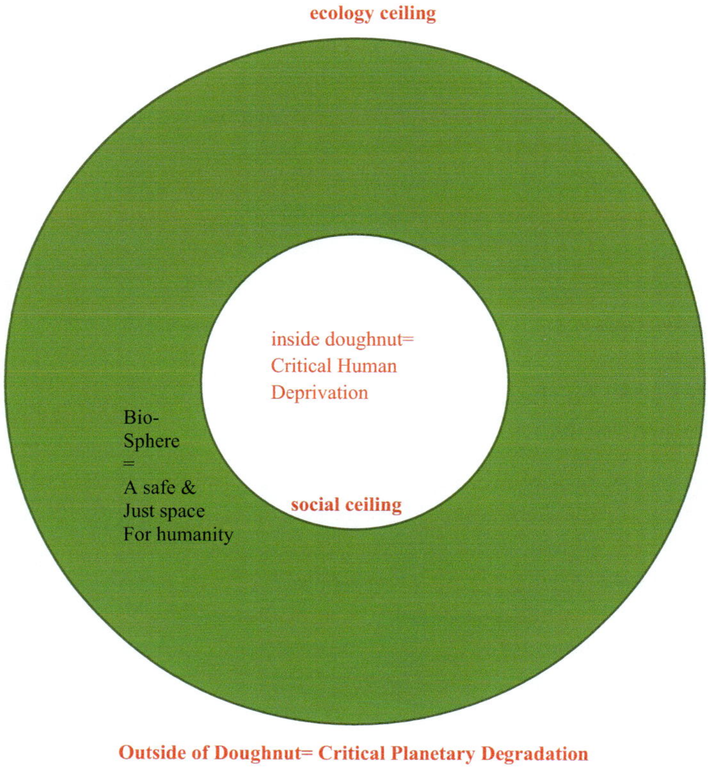

FIGURE I.2 Raworth's basic Doughnut diagram.

Raworth reminds us that the original meaning of economics is household management (*oikos nomos*), which in today's world means our duty of care to our planetary household, or planetary management (2017: 4). Why not, she then asks (43), focus on advancing the richness of human life rather than the richness of the economy (which after all is not a sentient being)? Drawing on the work of Earth-system scientists Johan Röckstrom and Will Steffen (entitled 'Planetary Boundaries: Exploring the Safe Operating Space for Humanity'[4]), Raworth presents us with a different economic model to the ubiquitous growth one, bringing together the two spheres of true economics which are, first, planet economics/ planet ecology and, second, social economics/social ecology (see Figure I.2). The first sphere encircles the second to produce what Raworth terms the biosphere: as we can see, in diagram form, sphere one – planet economics – encircles sphere two, the social economics, and the space between the two spheres creates the green biosphere which is Raworth's term for the ecology sphere, the space of planetary and human safety (11).

Raworth goes on to explain that, as we continue to pursue the growth model so, too, we increasingly transgress these two economic ceilings/boundaries which leads, in the first instance, to critical planetary degradation by overshooting the

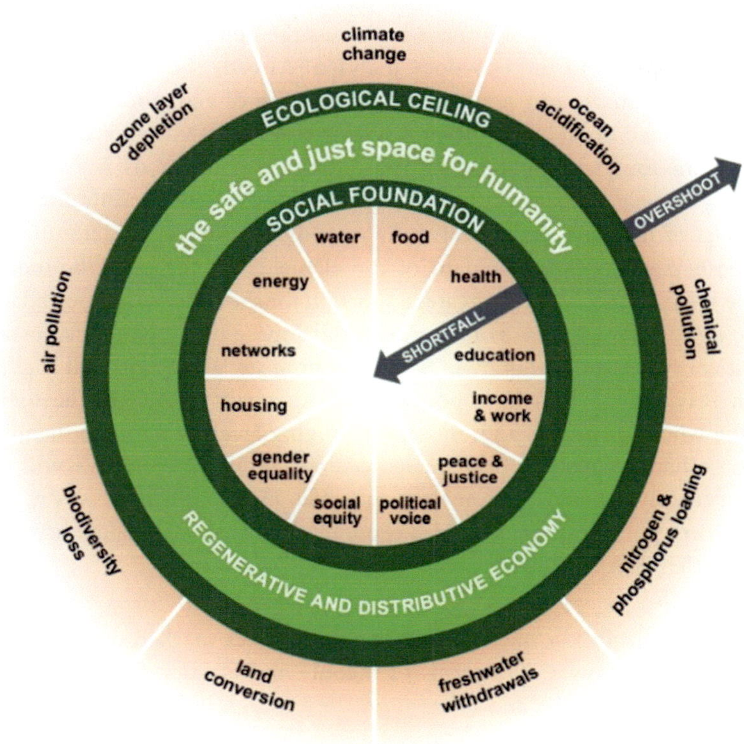

FIGURE I.3 Raworth's detailed Doughnut diagram (2017: 51).

ecological ceiling (written in red on the outer perimeter of the diagram), and, in the second, to a socially unequal and unjust space by falling short of minimum social standards, thus creating critical human deprivation by being in shortfall of the social ceiling (also written in red on the diagram). For its part, the biosphere revolves and recycles, regenerating planetary and social life, thus providing a safe and just space for humanity (44). In a more detailed diagram, Raworth expands on the present situation in which we find ourselves and in which evidence clearly points to how extensively the Anthropocene growth model has overshot and fallen short of minimum standards. See Figure I.3.

The ecological ceiling consists of nine planetary boundaries (all written on the outer perimeter), beyond which lie unacceptable environmental degradation and potential tipping points in the Earth systems.[5] The 12 dimensions of the social foundation (written on the inner perimeter) are derived from internationally agreed minimum social standards, as identified by the world's governments in 2015. Between the social and planetary boundaries lies the biosphere, an environmentally safe and socially just space in which humanity can thrive (Raworth, 44). But as you can see from all the red-orange heating, we have overshot and fallen short to a concerning degree.

As we progress through the four chapters of this book, on energy, pollution, agribusiness and migratory flow, I shall be pulling on this diagrammatic model and the essence of Raworth's argument to discuss the depth of the disaster confronting us as exposed in the eco-docs analysed. Already, however, thanks to this detailed diagram we can see how the one sphere of economics (planetary ecology) is impacted upon by the other (social ecology) and vice-versa. In brief: the abuse of one foundation entails the abuse within the other. As Ai Weiwei's documentary (*Human Flow*) makes clear, migration is the new condition of globalisation *and* of insidious ecological mismanagement. Indeed, the cumulative effect of poor planetary management of the elements on the outer boundary (what some are calling Ecocide[6]) creates a shortfall for the inner social boundary, thus inducing social inequalities of all orders, and a space in which peace, justice and a political voice are withdrawn. As the economists Wilkinson and Pickett (quoted in Raworth: 171–2) have pointed out, 'inequality damages the social fabric of the whole *of* society' and, moreover, 'higher levels of national inequality tend to go hand in hand with increased ecological degradation'. We will see instances of this in the eco-docs, not just in developing countries however, but also, as Raworth points out, in advanced economies where social inequality prevails, such as the USA:

> A study of all 50 US states found that those states marked out by larger inequalities of power – in terms of income and ethnicity – had weaker environmental policies and suffered greater ecological degradation.
>
> *(172)*

Indeed, as early as 2007, an interdisciplinary research group, headed by Dr Gregg Mikkleson, had established a connection between growing economic inequality and an increase in biodiversity loss. Here is what they found:

> 'While there is often a trade-off between economic growth and environmental quality,' says Mikkelson, 'this study suggests that there is a synergy between a different kind of economic development – namely, toward a more equitable distribution of wealth and the conservation of biological diversity.' For example, if the US were to achieve levels of income equality comparable to those of Sweden, the pattern reported in the research paper implies that 44% fewer plant and vertebrate species in the US would be in danger of extinction.[7]

The outcome of this research, according to Mikkleson, 'suggests that if we can learn to share economic resources more fairly with fellow members of our own species, it may help us to share ecological resources more fairly with other species'.[8] Social inequity is as much linked to ecological harm as it is to poor health, therefore, no matter the overall wealth of a nation.

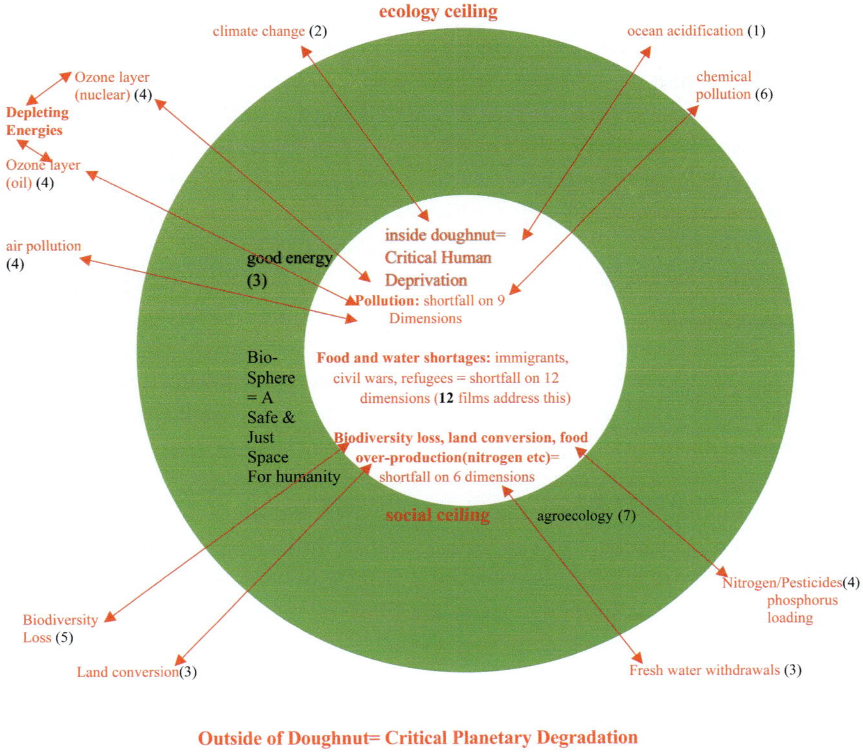

FIGURE I.4 Mapping the 57 eco-docs onto the Doughnut, a visualisation of planetary degradation and critical human deprivation.

As an initial guide to how we can use Raworth's theoretical model, let's now produce a visualisation of the 57 eco-docs' positioning in relation to the two boundaries (see Figure I.4).

And now let us briefly investigate what this initial mapping shows us. As we can see, these eco-docs demonstrate how, within our current Anthropocentric era, all nine of the planetary boundaries are overshot. Thirty-six of the 57 eco-docs directly address these transgressions of the ecology ceiling and their impact on the social ceiling, spelling out where the safe space for humanity is in shortfall. Thus, just to refer to the domain of agribusiness we can perceive its link to *biodiversity loss, land conversion, freshwater withdrawals* and *phosphorus loading (pesticides and fertilisers)*, for example. The result is that *water, food, health, peace and justice, political voice* and *social equality* are all in shortfall. Degradation of the soil through over-use, which is compensated for by fertilisation, plus deep-ploughing practices releasing harmful levels of CO_2 into the atmosphere have huge consequences on the ecology ceiling as much as they do upon our food chain and our ability to maintain a sustainable life which in turn impacts upon *social equality*. Furthermore, any attempt at protesting to the governing classes falls predominantly on deaf ears, driving parts of the world into civil disobedience, political disempowerment and, at its most severe, migratory flows. Twelve of the eco-docs, indicated in the inner circle of Critical Human Deprivation, have as their point of address the most deeply affected victims of these planetary transgressions, namely, internally displaced people, immigrants and refugees.

By applying Raworth's criteria we will also be able to see, when we come to analyse these eco-docs, how persistently the 12 dimensions of the social foundation are in shortfall. Key to this 'Doughnut Economics' approach is the immediacy with which we are able to visualise this deprivation and its sources. By *seeing* what is at stake we have the *knowledge* of what is at stake – affording us, thereby, a clarity that enables us to seek to change the way in which we choose to live on this planet. The drive for equity in all things begins with healing the Earth. And this is the theme of the seven agroecology and three energy films indicated in the biosphere (a safe and just place for humanity).

On a last note, it hasn't escaped your notice from the diagram, I'm sure, just how heavily tilted the current situation is towards overshoot and shortfall. But, as the following chapters will hopefully go on to show, we have both the means to know what's happening (the eco-docs) and the wherewithal, in the form of a regenerative economic model (the Doughnut), to understand how things are as they are, and indications as to how they undoubtedly could be, if humanity so chose.

Notes

1 Forms of living is a concept derived from Kathleen Millar's anthropological study of Brazil's *catadores* and refers to the pluriversalism of meaningful lives (2018: 9). See Chapter 2 where I discuss her concept in relation to *Waste Land*.
2 See: <https://almatyfilmfest.com/2019/06/08/the-undamaged-by-rozle-bregar-matic-oblak-miha-avgustin/> (accessed 7 April 2021).
3 See: Sophie Divry, 'Les Tartuffes de l'écologie, Pour repolitiser l'écologie' (Lyon, Parangon/Vs, 2007, 19–26); Charles Ruelle and Frédéric Neyrat <https://journals.openedition.org/labyrinthe/3762> and 'Le bouledogue rouge' <http://cequilfautdetruire.org/spip.php?article1520> (all accessed 2 April 2021).
4 See: J. Rockström et al., 2009 <http://www.ecologyandsociety.org/vol14/iss2/art32/> (accessed 28 April 2021).
5 This idea of the Earth's 'tipping point' now, in 2021, has a very real manifestation in the form of the Earth's axis shifting and moving the locations of the north and south poles as a result of ice-cap and glacier melting. See: <https://www.theguardian.com/environment/2021/may/06/how-melting-glaciers-have-accelerated-a-shift-in-earths-axis> (accessed 6 May 2021).
6 See the work of Ecocidal Defence Integrity, a movement founded by ecological lawyer, the late Polly Higgins.
7 The findings are quite clear: 'Mikkelson and his colleagues related indicators of income inequality and biodiversity loss on two different scales: among 45 countries worldwide, and among 45 states within the United States. They controlled for differences such as area and climate, human population size, and per capita consumption.

The same general trend is evident in both scales: societies with more unequal distributions of income experience greater loss of biodiversity'. In: Lisa Van Dusen <https://www.mcgill.ca/newsroom/channels/news/biodiversity-loss-linked-economic-inequality-worldwide-25398> (accessed 6 April 2021).

8 Ibid.

1

ENERGY – CRUDE AWAKENINGS

Extractive energy sources versus regenerative energy practices

It seems logical to start this study with ENERGY since the four planetary elements – earth, water, air, fire – which constitute the forces that drive Earth's existence are, paradoxically, in terms of the Anthropocene's exploitation of the Earth's resources, the very forces that are at the heart of global warming. In a mere 200 years we have managed, by tapping into the energy of the earth, to strip it of many of its inner resources which took millions of years to create. Take oil – 150 million years ago, extreme global warming caused the creation of oil. By contrast, in less than two centuries of oil exploitation we are nearing the end of its availability. By the early 1900s, the major producers and exporters were the USA, Venezuela and Russia; and as we know, the USA built its nation on oil. Yet, 80 years ago, half the world did not use oil. However, since the 1950s, we have come to use oil for virtually everything (from fertilising agriculture to greater mechanisation, from the many household objects we own, to the clothes we wear). Consider coal – first used commercially in China in 1000 BCE, it was not used massively until the nineteenth century, when the Industrial Revolution engendered an insatiable demand. In terms of its reserves, estimates suggest there is only a further 200 years' worth left. Both oil and coal cause air, soil and water pollution (as indeed do natural gas exploitation and fracking, new energy procedures of the late twentieth century). But not content with this and the pollution these industrial processes cause (responsible for 8.7 million deaths globally per year,[1] to say nothing of the impact on wildlife), we deforest our planet at a tremendous rate, thus endangering further the quality of the air we breathe. Indeed, with regard to the four planetary elements, because global warming has so seriously affected the ozone layer, these same forces have caused flooding in certain parts of the world, bush and wild-fires in some areas and desertification in others – this is the effect of our planetary plundering.

DOI: 10.4324/9781003130246-1

This apparent insatiability of human wants is the focus of all 13 eco-docs in this ENERGY category. All films pose the same question: continue as we are, or change? Should we remain addicted to fossil fuels and expand further our sourcing of fossil fuels and minerals, pursue uniformity in food production, build huge dams that will bring economic prosperity to a country at the expense of the natural environment, manufacture cheap energy whose disposal is unsafe? Or can we awaken from these dependencies which limit us in their destructiveness and open ourselves, instead, to the potential of more regenerative modes of economic existence in which caring for planet Earth is at the centre of our ethical nature?

In response to the above questions, 10 of the 13 eco-docs raise the thorny problem of economic need entrapped into limited expediency (see columns I and II in Table 1.1). And three (in column III) offer some working solutions constructed around regenerative principles.

The first four films (in column I) address the issue of nuclear energy: the bomb in *The Day the Sun Stood Still*; resistance to further atomic bomb testing as Greenpeace's opening gambit in *How to Change the World*; the storage of nuclear waste in *Into Eternity* and *Journey into the Safest Place on Earth*. The six listed in column II examine the impact of oil, hydropower and mineral prospecting for energetic ends, beginning with *Home*, which offers a broad sweep of the march of human progress over the planet. The other five investigate, respectively, oil dependency in *A Crude Awakening*; hydro-electric dams for aluminium processing in *Dreamland*, and hydropower dams for electricity in *Undamaged*; and abusive resource exploitation in Africa in *Virunga* and *Stealing Africa*. Finally, the three eco-docs in column III set out purposefully to demonstrate that change is a viable option: recycling waste, pursuing a non-throwaway ethos where food and plastic are concerned as with *The Gleaners and I*; exploring renewable energies in *The 4th Revolution*; training women in solar engineering as with *Solar Mamas*.

This chapter falls into three sections. The first focuses on a single eco-doc, *Home*. I do so because it provides the broadest sweep of the ecological issues at stake. Section 1.2 concentrates on eco-docs that investigate individual instances

TABLE 1.1 Table of energy eco-docs

I. Nuclear	*II. Oil/minerals/hydropower*	*III. Alternatives*
1. *The Day the Sun Fell* Switzerland (2015) 2. *How to Change the World* UK/Can (2015) 3. *Into Eternity* Den (2010) 4. *Journey to the Safest Place on Earth* Switzerland (2013)	5. *A Crude Awakening: The Oil Crash* Switzerland (2006) 6. *Dreamland* Iceland (2009) 7. *Home* Fr (2009) 8. *Stealing Africa* Den (2012) 9. *Undamaged* Slovenia (2019) 10. *Virunga* UK/Congo (2014)	11. *The 4th Revolution* Ger (2010) 12. *Les Glaneurs et la glaneuse/The Gleaners and I* Fr (2000) 13. *Solar Mamas* Den/UK/US (2012)

of the Anthropocene's relentless pursuit of non-renewable energy sources. The third section turns to the eco-docs that explore humankind's creative responses to the question of renewable energies and, with it, the concept of a bio-generative economy. Throughout, we bear in mind the nefarious effect of global capitalism and how it has fed into a seemingly endless insatiability of human *wants* addicted to the idea of growth and progress.[2] At the end of the second section I will provide a mapping on Raworth's Doughnut model so we can easily visualise what this means in terms of planetary degradation. Section 1.3, which offers solutions – both simple and complex – to our energy *needs* will also conclude with a Doughnut-mapping; this time however the picture that emerges is far more encouraging.

1.1 So let us begin – cards on the table: the four elements in chains and the performative exchange-value of *Home*

The eco-doc that most extensively documents the magnitude of this planet-asset stripping and despoliation is *Home* (Yann Arthus-Bertrand, 2009). It points in all directions to the effects of our commitment to energy growth over sustainability, as if the planet were a bottomless pit. Based around a consideration of the four planetary elements, and shot over 54 countries, this eco-doc focuses on the impact of the first two major revolutions brought about by the human species: Agriculture and Fossil Fuels – two revolutions that have become inexorably entwined. We learn how water and light brought plant life into being and, from there, how plant life altered the atmosphere and life forces began. Water, light, air, soil and trees created life. We then learn how humans first acquired the knowledge to harness those life forces, but then, over the millennia, subsequently abused that knowledge. This knowledge acquisition began once the human started to settle (10,000 years ago). Slowly but surely, the planetary balance (the Holocene) started to shift to the point of going out of kilter as it is now. First came the invention of agriculture, the use of beasts of burden for the heavy toil and our children as assets to work the land, a process which led to food surpluses and the concept of exchange-value. The major outcomes of this First Revolution were bigger settlements in the form of cities and civilisations and monetised exchange. The Second Revolution, much closer in time, and far speedier in its effect, was the exploitation of coal (beginning with the nineteenth-century Industrial Revolution) and then oil (in the early twentieth century) leading to massive industrial growth and a huge population increase – doubling under coal and growing by six times under oil. The effect: more people to feed, more cities to expand, rural exoduses and urban growth (often driven by a desire to survive rather than any expectation of new possibilities). To meet the needs of this population growth, *Home* informs us, a massive spur to the development of agriculture was required, which coal, then oil made possible. And since then Agribusiness has been on the march (as we will discuss in Chapter 3). We have stolen from the biodiversity of the Earth to grow crops to feed the livestock that feeds us. Thus, half the world's wetlands have been reclaimed to pastureland (or real-estate development, or flooding for hydropower – both of which we will see in *Dreamland*); similarly, around 43% of the world's rainforests have been forfeited to produce soya and palm oil. Trees, our oxygen suppliers, have been turned into meat and other food products.[3] Wetlands, the most biologically diverse of all ecosystems (purifiers of water and carbon capturers amongst other properties) are sacrificed to human wants.[4]

Home is constructed like an opera. Wagnerian in its dimensions, the magnitude of the planetary destruction appears matched, in production-value terms, by the film's own excesses. With its fusion of music, voice-over narration and painterly images, it could be argued it out-performs itself. A massive operation of visuals and sound, exposing the delusional beauty of our planetary plundering, *Home* is a hymn of the sublime addressed to the non-sublime. Our senses are bombarded. *Home* is a Blakean visual-sublime of awe and terror; a monumental partition devised from musical influences from all corners of the Earth; a censorious voice-over endlessly intoning our disregard of the evidence in front of us. First, there are the extravagant gyro-technics of the film's shooting – the director, Yann Arthus-Bertrand shot this film mostly using aerial shots taken from a small helicopter and using high-definition Cineflex cameras suspended from a gyro-stabilised sphere attached to rails on the base of the helicopter (Duvall, 2017: 63).[5] Second, Armand Amar's powerful, sometimes throbbing, sometimes sweepingly lyrical score – made up of 24 tracks, drawing on all varieties of music, and thematically linked to all the elements present in the film – is a relentless reminder of the universality of our desecration.[6] Third and finally, in the English version, Glenn Close's beautiful voice unrelentingly remonstrates with us as we go 'faster and faster' to our self-induced perdition (much as the mellifluous Mephistopheles did with Faust in Gounod's opera). The whole effect of this painterly aestheticised apocalyptic text (with a ten-minute coda of hope at the end) is overwhelming. Given this film was produced by Luc Besson and his company EuropaCorp, this extraordinary global extravaganza should not surprise,[7] and indeed it delighted spectators – reaching well over 50 million *since* its one-week free release in June 2009.[8]

The controversy over the film's $12 million budget, sponsored by the Qatar Foundation and primarily Kering (an international group based in Paris specialising, in 2009, in retail and luxury goods) has already been mentioned in the Introduction. And it seems easy to point out that, much as with the film's production context, there is also an uncomfortable tension within the film itself, this time between the *aestheticisation* (the sublime), through sound and image, of what we are shown and the *content* and *meaning* of what we see (the non-sublime). By this I mean that just as there is a problematic funding context so, too, there is a sense of glut within the film itself (perhaps an effect of the 488 hours of footage the editors had to deal with to bring it in at 90 minutes). Putting everything in, packing it high with so much verbal information and sonorous flow, leaves very little space to reflect, let alone respond (it leaves one truly breathless). Almost as if the technological feat of bringing all three documentary parts (image, music and narration) together and in on time (for 5 June 2009, World Environmental Day) was as significant as the film itself – so, the PRODUCT, the spectacle itself becomes *that* which MATTERS. Furthermore, the visualisation of the damage done to the planet is powerful, but it is also intentionally shaming.[9] These are paintings of the Rape of Gaia (in a Blakean sense) and they are meant to shock, doubtless, but there is no human scale to attach to, apart (perhaps) from the brief mention, at the end, of examples of good practice and slogans such as 'make harmony with nature the rule' or 'agriculture on a human scale' and calls to be a responsible consumer, all of which are intended to reassure us that it is not too late. Thus, the risk this block-blustering film runs is that we remain, on the one hand, simultaneously shamed and overwhelmed by the enormity of our misdemeanours and, on the other, perplexed rather than empowered as to how to act for change, to save our *Home*. That being said, there appears a tangible argument, however, for the film's *re*formative exchange-value in that Kering, since 2012, has committed itself to an environmental plan targeting a 40% reduction of its global environmental impact by 2025.[10]

Indeed, if we slow down our own response to this film and consider it alongside Raworth's Doughnut diagram, a somewhat different or nuanced way of evaluating this eco-doc occurs. In terms of performative exchange, *Home* has great value in its visualising our transgressions of the two planetary boundaries (ecological and social). Insofar as the painterly images, in their own excessive sublime, directly reflect our own addiction to the ugly beauty of fossil fuels, we are exposed to what it means to take. These images now become images of the degree of our insatiable greed, and the consequential damage to the environment of those from whom we take (indigenous peoples being a primary example, biodiversity the other). *Home* exposes how all nine of the planetary boundaries are transgressed by our Anthropocenic plundering. And, because of the way in which powerful nations and multinationals *negotiate* this plundering – whereby, just to cite one example, half of the world's population, despite living in resource-rich countries, nonetheless live in poverty (a point we shall return to) – most of the 12 dimensions of the social foundation are frequently in shortfall.

Home also exposes the underlying problem of our refusal to slow down. Because capitalism is so committed to the economic-growth model and is thereby slavishly attached to the principle of GDP growth, we precipitate ourselves faster and faster and pursue farther and farther afield in order to obtain more and more. As Raworth (2017: 261) explains, 'the last two centuries of extraordinary growth in high-income countries is largely due to the availability of cheap fossil fuels'.

It is the 'cheap' of this energy which is the primary reason behind big energy corporations' refusal to consider turning their attention to renewable energy. Because the fossil fuels still remain (in relative terms) an inexpensive commodity, energy companies can continue to make huge profits and satisfy both their shareholders and governments by contributing to the nations' GDP. Thus, financially, politically and socially, high-income countries remain addicted to growth (see

Raworth: 272–80). But as we know the planet has limited resources; it can only provide us with what we need in a cyclical and regenerative manner. Continuous cumulative growth (which is how GDP works) is thus brought face to face with the carrying capacity of the Earth. And the contradiction is clear: money cannot endlessly accumulate money, any more than fossil fuels can be forever available. When fossil fuels run out, economic growth will decrease. Clearly, then, the capitalist growth model has to change. As Raworth indicates, we need to transform the economy 'from ever-growing on an unstable trajectory to ever-oscillating within a stable range' (271) – an economy that is regenerative and distributive by design (267). But financiers and governments resist this idea of 'sectorial transformations' because over the past 200 years 'capitalist economies have restructured their laws, institutions, policies and values so they are geared to *expect*, *demand* and *depend* upon continual GDP growth' (267 Raworth's emphasis). This position appears, however, to fly in the face of evidence since, over the past 40 years, GDP's slow-growth periods outstrip the moments of fast growth (currently many economies are flat-lined with zero or near zero interest rates and stagnant wages, so the evidence is here for all to see). Yet rather than respond to this truth, governments deregulate, cut red-tape and privatise, which in turn causes 'speculative bubbles, home price increases and debt crises' (269).[11]

On this issue of non-renewable resources' extraction and production, which also includes the production of Nuclear energy – let us now turn to the other eco-docs listed in columns I and II. More measured than *Home* in the tone of their investigation into the effects of linear thought and practice (which is endemic in growth economics), and certainly more conscious of minimising their own environmental footprint, these films explore the *how* of planetary despoliation, and expose as its *source* the mental addiction to certainty held by the elites' boundless faith in the neoliberal way of global capitalism – an unsurprising addiction, doubtless, since they represent the 2% of the population that owns 50% of the world's wealth and resources.

1.2 Cheap Energies – an unholy triumvirate of the Clean, the Bad and the Dirty: Nuclear, Hydropower, Oil

This might strike you as a peculiar triumvirate of 'bad' energy sources. Surely not all three of them are harmful to the environment, I hear you say. Hydropower and Nuclear energy are supposedly clean energies, so we have been told. However, there are huge costs attached. Some, as with Nuclear, we never see because radiation is invisible (until the effects are manifest months, years or decades later). Nor indeed do we have any idea of the eventual impact of nuclear-waste disposal. Others, such as Hydropower are a natural source for producing electricity – and kinetic run-of-the-river energy, or natural waterfalls are indeed examples of environmentally friendly energy producers. Less so dams and reservoirs, both of which have a significant environmental impact including ruining river ecosystems, flooding land and causing greenhouse gas emissions from rotting vegetation below in the water, depriving inhabitants of their homeland and affecting their livelihood.[12]

Both Hydro and Nuclear energy are forces which, whilst we may delude ourselves into believing we control them, nonetheless are a power unto themselves. Dams can break. The cheap energy they can produce for hydro-electric and geothermal plants can bring in its wake polluting industries (as we witness in *Dreamland*). Nuclear energy has produced bombs of mass extinction (and we still live in fear of countries who own this technology). The Nuclear industry, in creating power plants across the world, has generated security problems of its own (Chernobyl and Fukushima being obvious examples). And then of course there is the nuclear-waste problem, often shrouded in mystery. Sellafield in the UK keeps this waste in 'high-integrity stores'. The USA speaks of storing it in the Yucca Mountain, and has been speaking of doing so ever since 1987;[13] meanwhile it stores waste either in facilities where it is generated or in landfills or in the oceans in the Arctic area.[14] Finland is burying its waste deep into island rock (see *Into Eternity*). Japan plans to bury theirs too, but currently holds their nuclear waste in storage pools above ground; neither of these procedures bode well for a country that is vulnerable to earthquakes and tsunamis.[15] Indeed, in 2011, the earthquake that hit the Fukushima Daiichi nuclear plant caused it to release thousands of tons of radioactive water into the Pacific Ocean (see *The Day the Sun Fell*).[16] Given that High-Level Waste can last for up to 100,000 years, the leak-ability of this dangerous material should cause us all concern. As the writer Arundhati Roy, in relation to these two energy systems, puts it so eloquently in her 1999 blog *The Greater Common Good*:

> Big Dams are to a Nation's 'Development' what Nuclear Bombs are to its Military Arsenal. They're both weapons of mass destruction. They're both weapons Governments use to control their own people. Both Twentieth-Century emblems that mark a point in time when human intelligence has outstripped its own instinct for survival. They're both malignant indications of civilisation turning in on itself. They represent the severing of the link – the *understanding* – between human beings and the planet they live on.[17]

Neither oil nor nuclear energy are unlimited in their availability. And energy scientists have been sending warning signals for a considerable time now, as *A Crude Awakening: The Oil Crash* explains. In 1955, the geologist Dr Marion King Hubbert predicted that petroleum production of oil reserves would resemble a bell curve and that in the USA they would peak between 1965 and 1970. Although ridiculed at the time, his theory, the Hubbert Peak Theory, proved to be correct. Moreover, predictions for all oil-producing countries reserves are following the same bell-curve model – currently, in global terms, we have passed the peak years' range (2000–15). Similarly, for nuclear energy. Scientists interviewed in *Into Eternity* explain that nuclear is a short-term solution to energy needs because the uranium-235 fissile isotope (a high-grade nuclear power reactor) is a finite non-renewable resource. Both the OECD Nuclear Energy Agency and the International Atomic Energy Agency are agreed that the uranium peak was in 1980. Until 1990, uranium production was in excess of the consumption required by power plants, but since 1990 that trend has been reversed. As the world's population expands, so too the demand for electricity. China and India, who together make the world's largest energy-consuming class, are expanding their nuclear-plant capacity, adding to concerns that the former prediction that there is 70–80 more years' worth of uranium-235 will prove incorrect. The revised thinking is that there is perhaps only 40–50 years available.[18] We already know that a lack of resources is a magnet for wars. We have witnessed how the struggle for oil took off in earnest in the 1970s and conflict (often armed) has continued to intensify.[19] To date, uranium has seen a different kind of conflict. There are tensions between dominant nations and countries endeavouring to develop their nuclear-power capacity. The USA–Iran situation is a case in point (with the USA not trusting that Iran won't develop their nuclear capacity along military lines). More typically, it would appear that uranium ventures are more likely to increase conflict at the intrastate level where countries are engaged in civil or factional conflict. These, at least, are the findings in two of the African producing nations, the Democratic Republic of Congo (DRC) and Niger.[20]

Whilst we note there is still a considerable amount of mystification occurring around oil resources, strangely most countries are fairly open about the reality of the terminal decline of uranium, perhaps because other forms such as fusion-reaction are being explored. It is also the case that several nations are turning away from nuclear plants (Belgium, Germany, Spain and Switzerland in Europe for example), or refusing to allow them to be built (Italy is a leader in this). One country that most definitely is not, however, is Japan which, in 2012, reversed its earlier policy (following the Fukushima disaster) to reduce the country's reliance on nuclear energy, and since 2014 has committed instead to restart plants and pursue its building plans for new plants.[21]

Let us, first, consider briefly the Nuclear and look at just two of the eco-docs listed in column I, *The Day the Sun Fell* and *Into Eternity*. For they are exemplary in raising important questions as to how humankind deals with the living reality of this ultimate tool of power and destruction which is all of our own scientific making. Two key elements emerge, memory and silencing. The desire to forget and the refusal to be silenced dominate the structure of *The Day the Sun Fell*. The desire to consign the dangerous waste product of this fearful power to oblivion is the running thread of *Into Eternity*.

Into Eternity (Michael Madsen, 2010) explores that point where, as Arundhati Roy indicated above, human intelligence has outstripped its own instinct for survival. Or as David Ratcliffe puts it, the moment when 'humans have mastered the power of the universe' only to realise that 'their discovery has outstripped human ability to understand the consequences'.[22] This uncertainty is echoed, throughout this eco-doc, by the scientists involved in the Onkalo repository project, Finland, which is intended to offer a permanent solution to nuclear-waste disposal. The first half of the film focuses on the function of the project, burying the waste 500 metres and even further down into the Finnish bedrock (1.8 billion years old) – the only safe place for waste to be stored (as opposed to on land or in the ocean, so the scientists affirm). The second half speculates on the future of such sealed and concealed repositories remaining safe and secure from human intrusion for the next 100,000 years (the time estimated by these scientists for the waste to lose its toxicity – the Americans, we are told, quote a million years, to be on the safe side). *Into Eternity* exposes the absurdity of this speculation. How is it possible to predict what human form might be around by then; what about the stability of the storing towers and how to guarantee that every generation is reminded of the necessity to 'remember forever to forget' (to quote the narrating director Michael Madsen)? And what of the massive costs involved in drilling these hugely cavernous spaces of oblivion? If the nuclear energy path had not been pursued, that money surely could have been invested in renewables. All these issues raised by the eco-doc serve as painful mirrors to the folly of mankind. The project Onkalo, meaning hiding place, could hardly be better named – if this is to be the 'longest lasting remains of our civilisation' (again Madsen), perhaps it was wisest to conceal it. For, if discovered, it will obliterate the planet. As Madsen asks 'if we cannot understand it or grasp it, it's impossible to act responsibly'. The closing aria to the eco-doc by Edgar Varèse (based on Paul Verlaine's poem, *Un grand sommeil noir,* which includes these poignant words: 'let all hope sleep ... deep in a cavern, silence, silence!') resumes the chilling awe-fulness of this predicament we have landed ourselves in. And yet the Atomic bombing of Hiroshima and Nagasaki certainly forewarned us – that by possessing the powers of the universe we simultaneously have in our hands the ultimate tool of our own destruction.

The initial intention behind Aya Domenig's documentary *The Day the Sun Fell* (2015) was to investigate her deceased grandfather, Shigeru Doi's experience of the aftermath of the 1945 Hiroshima atomic bombing, during which time he was a doctor of internal medicine at the Hiroshima Red Cross Hospital. In 2010, Domenig set about recording the testimonies of other medical staff who also survived the bombing effects, a doctor, Shuntaro Hida, and a handful of nurses as well as her grandmother (also a nurse, but not serving at the time). Only Dr Hida and one nurse, Chizuko Uchida, actively bear witness by giving talks in schools all around the country. The three other nurses can't speak of it, adding 'the bomb destroyed us'. What emerges at this stage of Domenig's investigation are the lengths to which both the American occupiers and the Japanese authorities were willing to go to keep the true impact of this nuclear attack hidden. This included silencing the victims and the doctors attending the patients (they were not allowed to keep records) and confiscating filmed footage by Tokyo journalists (not made public until 1958[23]). The effects of the bomb's radiation remained a military secret, most particularly the evidence – noted by medical staff at the time – of internal radiation (patients' bodies were rotting from the inside, bleeding from all orifices). Very few people who were actually in Hiroshima at the time survived, but outlying areas were also deeply affected and people succumbed to all sorts of health problems. As Uchida states 'radiation destroys every organ it can. Even though it's invisible, it gets into the body and things decay'. To which Dr Hida adds 'you were afraid of getting ill, doctors collapsed and died, but the Americans claimed there was nothing wrong'. According to Hida, to date the USA still insists internal radiation does not occur. This veil of silence was maintained by survivors afraid that it would make it almost impossible for their children to get married. Even Domenig's grandfather said he could not put into words what he experienced.

Barely one year into making the documentary, the 2011 Fukushima disaster occurred and took Domenig's documentary into the second territory of the Nuclear's power – as energy provider, but with equally risky consequences to people and the environment once its radiation leaks out into the atmosphere. Both the plant operator, Tokyo Electric Power Company (TEPCO), and the government in the form of the Ministry of Economy, Trade and Industry sought to minimise the real impact of this disaster by reassuring people. When you consider that 30% of Japan's electricity is produced by nuclear plants, it is obvious the economic fall-out, to say nothing of the power outage, would be considerable, so it behoves officials to sell the public myth of 'absolute safety' where nuclear energy is concerned. Families that had initially been evacuated returned to their homes in the belief that they wouldn't come to harm, that their children could not be affected and that, anyway, their health insurance would cover any health problems.[24] Hido and Uchida explain their outrage at the government's response – for them it is the same set of lies once more. To make claims as did the government, in August 2011, around the period of the Hiroshima Peace Day, that the danger had passed and to then attend the 6th August ceremony in remembrance of the A-bomb victims, strikes Hido as the ultimate hypocrisy. Hido remarks, 'the government does not want the truth to come out and to have to pay compensation'. Uchida expresses this hypocrisy even more poignantly,

> even though radiation is harmful, the Japanese government does not want to give up on Nuclear. Radioactivity was used for murder in the war, but using it after the war also puts us on the path of death. But plutonium makes our country powerful so they won't relinquish it.

The same institutional rhetoric used post-Hiroshima-Nagasaki is present once more; 'there's no danger', people are told, 'no internal radiation risks' – to which Hido responds that radiation once out in the environment will inevitably pervade the atmosphere, enter the food cycle; besides, radiation sickness can develop over many years (as we witness in the case of Domenig's grandfather who took 30 years to die and, presently, her grandmother who is slowly dying of cancer). 'The same will be true for Fukushima', Hida says, 'the government will try and dispute it'. He continues, 'Japanese culture is very patriarchal and needs to listen to the women who know better what is going on, because they are the carers. Men and women need to join forces to protect the future'. As environmentalist and physicist Amory Lovins further explains:

> Japan's rigid bureaucratic structures, reluctance to send bad news upwards, need to save face, weak development of policy alternatives, eagerness to preserve nuclear power's public acceptance, and politically fragile government, along with TEPCO's very hierarchical management culture, also contributed to the way the accident unfolded. Moreover, the information Japanese people receive about nuclear energy and its alternatives has long been tightly controlled by both TEPCO and the government.[25]

For all these reasons, both Hido and Uchida, despite their great age (in their 90s), feel compelled to keep on speaking out, fighting for the truth to be acknowledged. Of the government and officials, Uchida has this to say, 'these people will continue to mislead us, I must speak out'. She is right. In the immediacy of the disaster, TEPCO was ordered by the government not to use alarming terms, such as 'reactor meltdown', to describe the impact of the tsunami on the power plant. Uchida tells us how at the time of the disaster the officials made mistakes in evacuating people because, again, the establishment played down the risks and suppressed information about the radioactive plume. As a result of this misinformation some people were evacuated from lightly affected areas to more heavily impacted ones. Knowing the dangers, we see Uchida taking action by inviting a mother and child to shelter in her home until the worst is passed and teaching her the techniques of natural healing methods which she has used ever since the Hiroshima bombing.

At the time of the disaster, large amounts of water contaminated with radioactive isotopes were knowingly released into the Pacific Ocean, the effects of which were still being recorded in 2013. This led to an unsuccessful attempt at a purifying process within the Fukushima plant (and, as of April 2021, the Japanese government declared it was proposing to release over one million tons of this waste-water into the ocean, because its cisterns were saturated). Meanwhile, 18 months after the disaster, the Japanese government announced plans to make Japan nuclear-free. However, a change of government three months later led to a reversal of the decision and existing reactors, far from closing down, were restarted. To compensate for the loss of power any new nuclear reactors would have brought, plans to implement coal power plants were revived. Environmentally speaking, Japan has lurched from one ecosystemic disaster to another.

Both *Into Eternity* and *The Day the Sun Fell* make clear that, even though we have the awareness that we have manufactured an energy force that we cannot control, nonetheless, authorities will do all in their power to conceal this truth. Some nations pride themselves on being a nuclear power – and won't relinquish it. America certainly, and Japan. The former knows it has first strike capability (atoms for war). The latter sees, in its huge reserves of plutonium, evidence of its powerful status. Nuclear not only allows for its use in nuclear reactors (atoms for peace), it also makes possible the potential to quickly produce a nuclear weapons arsenal (something Japan has foresworn of course, but just having the potential is a sign of power).[26]

Even if Nuclear is promoted as a clean ENERGY, nonetheless it is a producer of potentially colossal after-effects – much of which is invisible (which doesn't mean clean) until it is too late. The oil industry, for its part, however, with its huge political and economic engine, often shrouded in murky dealings, coupled with its detrimental impact on the environment, is a very present and visible danger and is certainly far from clean. Moreover, what emerges from a study of the six eco-docs in the 'Oils/Hydropower/Minerals' category (see Table 1.1, column II) is the compulsive pursuit of the One Big Idea concept which, in this context, boils down to CHEAP ENERGY.

A Crude Awakening: The Oil Crash (Basil Gelpke & Ray McCormack, 2006) makes clear the tentacular nature and cost of securing oil. Since 1945, the eco-doc informs us, the basis of US foreign policy has been to secure *cheap* oil in exchange for protecting rulers of Saudi Arabia who trade oil for arms. Furthermore, in recent years, the wars conducted in Iraq by the two Bush administrations were about securing access to oil (and much less about ideology). Arms for oil also helped fuel the South Sudan Civil War (2013–20). This time a cabal of ruling South Sudanese elites were propped up by a complex web of corruption that included American arms dealers, British business tycoons and Chinese oil giants.[27] Not only, then, is the energy itself a dirty energy, so too are many of the practices involved to secure it (often disguised as foreign policy). And the impact of these wars and civil conflicts on the various nations' peoples, or indeed ethnicities, is death in the hundreds of thousands, displacement and migration of refugees in their millions.[28]

In terms of the 'dirty' of this product, mention must also be made of the numerous oil spills, one of the more recent ones being the 2010 BP Deepwater Horizon disaster in the Gulf of Mexico (which I discuss in *Film Ecology*, 2010: 18–19).[29] But, what I now want to turn our attention to is the adjunctive 'dirtiness' of this product, namely, the financial system that relies so heavily on oil. As *A Crude Awakening: The Oil Crash* points out, the financial system is fuelled by petrol dollars. The world's 60 biggest banks have provided $38 trillion of financing for fossil fuel companies since the Paris Agreement of 2015. Equity markets continue to invest (fossil fuels' stocks largely finance pensions); currently, 31/100 FTSE companies are well above what's needed to hit the Paris Agreement target, and so on.[30] There is, of course, a huge danger attached to this financial speculation system, because oil companies will be over-valued when oil runs out, at which point the economy will face the prospect of another Great Depression. And for the simple reason that we are living on more oil than is being produced. However, in order to keep the price of oil stable and protect both their production quota (determined by OPEC which controls 75% of the world's crude oil reserves) and their financial standing, oil-producing nations and companies quite regularly misrepresent the true amount of oil in their reserves (sometimes keeping the estimate low so as to increase the prices, as occurred in the 1970s, sometimes upping production in the light of lowered prices so as to guarantee the same income). Thus, I'll cite one example which drew disastrous consequences for the country in question, Kuwait. Throughout the 1980s, when oil prices were at a much lower level compared to the 1970s, Kuwait's oil production was considerably above the OPEC quota (around 50%). These high production levels kept revenues down for Iraq – itself low on reserves and in debt to Kuwait who helped fund Iraq during the Iraq–Iran war 1980–8. In 1989, Kuwait refused to write off the debt. This, in turn, led to Saddam Hussein's invasion of Kuwait in 1990 and annexing it (and of course its oil fields) within a matter of hours. And we know what happened next.

Presently, in this first quarter of the twenty-first century, there is not the oil glut of before, and OPEC is well aware of this. This intergovernmental regulatory body, whose purported role it is to stabilise the oil markets, has shown itself incapable of doing so. Unsurprisingly, given that OPEC's most powerful member is also its largest crude-oil producer, Saudi Arabia.[31] This is a nation that regularly flouts the quotas, arguing that as a high-producing country it should have the lion-share of the market and that, as with capitalist practices, if they can afford to sell their oil at a low price, then so be it. Poorer-producing countries in OPEC, pushing for production cuts (which would allow for the price of oil to rise), are up against the Saudis' concern that to drive up the price of oil would drive industrial nations to develop alternative fuels (shale gas in the USA being one example). And so their policy is to meet demand. Meantime, the public face of OPEC isn't immune to its own mis-demeanours. As one of the interviewees in *A Crude Awakening: The Oil Crash*, the geologist Dr Colin Campbell, puts it, OPEC practices its own misrepresentation of data. Inflated figures are put forward, suggesting more oil is there for the taking than is the case. And as the eco-doc makes clear, the numbers just don't add up. But then, as Campbell points out, the OPEC countries do not care about what happens 20 years from now, when oil will virtually have run out. What matters is wealth in the present – with global revenues for oil at $2.7 trillion in 2019, this is hardly the time to expect oil-rich countries to move towards developing alternative energies.[32] Despite the impact of the coronavirus pandemic upon the oil industry which has reduced profits by a trillion dollars, there are few signs that oil companies or insurers are shifting their thinking.[33] Only the insurers Aviva are committing to a zero carbon footprint by 2040. BP and Shell have continued to defer any real commitment to renewables; governments are planning fossil fuel production increases of 2%; moreover, G20 countries are giving 50% more toward coronavirus recovery to fossil fuel companies than to clean energy.[34]

This compulsive pursuit of the ONE IDEA and the need to resource it becomes the single driving force behind appropriation. And, once obtained, a pretty monolithic culture is installed whereby everything depends on that ONE source (be it oil, minerals for heat- and electricity-conduction or hydro) and who controls it. This driving force produces new kinds of 'colonising' or 'occupation'. Thus, typically, countries that are in economic need but which are rich in energetic resources become easy prey to giant corporations looking to expand their operations, and to that effect employ what the economist John Perkins calls Economic Hit Men (EHM) to carry out the work.[35] This same John Perkins, interviewed in *Dreamland*, explains the process:

> It is fair to say that we EHM have created a world which is a first truly global empire ... We work in many different ways, but most commonly we identify a Third World Country – most typically it's a Third World country – whose resources we covet and then we'll arrange a huge loan to that country. However, most of the money doesn't ever go to the country. Instead, it comes from organisations like the World Bank and goes directly to one of our own corporations which will build big infrastructure projects in that country, things like hydroelectric dams and industrial parks, big industries, things that do not benefit the majority of the people at all but help a few very wealthy people, big corporations most of all. In the end, the country is left holding a huge debt. Such a huge debt, they can't possibly repay it. So at some point we go back, tell them: 'you owe us a lot of money which you can't pay back ... so sell your resources real cheap to our corporations'.

In four eco-docs we see precise examples of this practice. Two – *Virunga* and *Stealing Africa* – occur in so-called Third World countries (to use Perkins' term), respectively, the DRC and Zambia (more on these two films later). But it isn't just identifiably poor countries that get exploited; countries with struggling economies elsewhere also get hit. In the other two eco-docs, the targeted countries are European. *Undamaged* exposes the corruption of politicians selling out their rivers, along the Balkan Peninsula, to foreign hydropower developers and the ensuing damage-impact to local people and the environment. *Dreamland* takes place in Iceland – a European country which, since the withdrawal of Americans from their military base, beginning in the mid-2000s, has suffered several waves of economic crises (including the bank crisis of 2008). Since the 1990s, however, thanks to the availability of geothermal power and the harnessing of its many rivers and waterfalls for hydro-electricity, Iceland has become almost entirely resourced by renewable energy (99%).[36] So much so that its electric power reserves are in excess of need. When the various crises of the new millennium hit, therefore, the Icelandic government, fearing a rise in unemployment, set about vaunting its cheap power to attract foreign industries. In stepped the American aluminium company ALCOA.

Dreamland (Andri Snaer Magnason, 2009) reveals how Iceland became a sitting duck for ALCOA's Economic Hit Men. Politicians, mayors and a large majority of the Icelandic population were totally behind the scheme to build huge dams to enable the establishment of several aluminium smelter plants across the island. On the Icelandic side, politicians promulgated the fear rhetoric of economic depression and job uncertainty to persuade the islanders. We also learn how other elites looked to see what profit they could derive, others still to the prospect of lucrative jobs on some board or other after their term in political office. On the Americans' side, profit-driven ALCOA demanded the construction of an even bigger hydro-centre, the controversial Kárahnjúkar Hydropower Plant which entailed the building of five dams and three reservoirs at the expense of flooding huge areas of natural wildlife conservation areas. Meantime, the impact of the construction of ALCOA's smelting industrial plant, some 50 miles to the east, also had its devastating effect. Agricultural land was appropriated from farmers; 'it was easy to persuade people who have so little that the aluminium plant was a gift' the film tells us. Worse still, the neighbouring cropland of working farms became transformed into a dilution zone by the polluting airflow from the plant. The affected air also produced breathing problems. Dust emanating from the smelting plant blocked views; even airlines complained about visibility. The flooding has caused a huge decrease in bird species. Sediment running off the holding basin created by the construction of the dam has been responsible for the widespread deterioration of Lake Lagarfljót's ecosystem.[37] Promises made by ALCOA to the Minister of the Environment to build local amenities have so far come to nought. Instead, towns near the plants have been built, creating a boom in the real-estate market (mostly unaffordable to local islanders). Further, ALCOA is so big it simply pushes aside the possibility of other companies setting up businesses or building new companies. And, as to the promise of jobs for the islanders, cheap labour has been bought in from outside (partly in reaction to Iceland's very strong trade-union tradition, protecting wages, doubtless) – so employment is not as plentiful as first promised by the government.

Because of Iceland's low energy prices, ALCOA saves $200m per year (the annual salary, we are told, of 7000 teachers). This saving represents a third of what Europe and the USA pay. Initially, however, the project brings prosperity, at least to the Icelandic banks and the government. To finance the new constructions, the banks are obliged to increase interest rates. This, in turn, allows the government to take a bigger slice of the GDP. Sadly, it also means there is very little benefit to the local population from whom, according to this eco-doc, the precise energy price is kept secret. Finally, as the Kárahnjúkar project reaches its maximum power, to all evidence Iceland's GDP will first stagnate and then begin to shrink and so another big project will be needed. And the signs are that ALCOA is already poised. Aluminium production is extremely energy consuming, so it is hard not to believe ALCOA won't press for more plants to further its financial growth. Politicians tried to persuade the islanders they had found the sure solution. *Dreamland*, however, exposes the folly of binding a nation's well-being to ONE IDEA, ONE METAL, ONE COMPANY, ONE DEAL – hardly 'a dream come true' as the Environmental Minister, Siv Friöleifsdóttir, puts it, she who *reversed* the National Planning Agency Commission's refusal to allow the dam to be built.[38] From its former 50-year dependency for economic prosperity on America's military presence during the Cold War period, Iceland finds itself again economically dependent on an American presence, this time the giant ALCOA – which of course has strong links with the American military.[39] The irony in the eco-doc's title – *Dreamland/Draumalandiö* – escapes none of us.[40]

Miha Augustin's *Undamaged* (2019) follows a group of kayakers (the Balkans River Defence Tour) who travel down 18 wild rivers from Slovenia to Albania, to expose the damage already incurred by dam-building and to protest against money-grabbing developers whose proposals to construct yet more dams are well underway (the plans are for upwards of 2500 dams, 113 inside national parks). This journey through five countries documents how previous dam schemes (from the 1950s onwards) have wrought untold damage upon both the environmental ecosystems and the local inhabitants. In the post-communist era, from the 1990s onwards, energy corporations from outside, the Austrian-German Kelag dam construction company in this instance, having successfully persuaded local government of the benefits a hydro-plant will bring to their community, have then proceeded to fall short on their commitments for job creation and – where a valley flooding of a village occurs – new and secure housing. The reality, as *Undamaged* makes clear, is that land is bought off the local people for peanuts, the houses they are moved into are well below standard, jobs do not materialise so the younger generations have to move on to find work. Moreover, these village displacements can mean the breaking up of communities, so the social consequences are significant. The impact on biodiversity is equally extreme. Trees underwater in the flooded valleys eventually decompose and release methane. Fish can no longer swim upstream to reproduce (at one point we see how the dam in Slovenia blocks the fish from rising up to their spawn areas and, worse still, how its twice-daily flood-water release actually kills the fish). As the leader of the kayakers Rok Rozman says, 'having already killed off their own rivers, the Austria-German Kelag company has come to the Balkans to bring electricity back to their own homes'. If the projected 2500-odd dams are built, as the Balkan River Defence organisation makes clear, 'thousands of kilometres of wild river will be destroyed with irreversible damage to ecosystems and endemic species populations'.[41]

Stealing Africa and *Virunga* are further distressing witness documents to the disturbing practices of wealthy giant corporations – this time resource extraction in Africa. The dynamic is the same; rich prospecting companies come offering inducements to political leaders who, in turn, fail to protect the resource-wealth of their nation and either turn a blind eye to the international company's fiscal fraud (in the form of tax avoidance and tax havens) that otherwise would enrich their country (as in *Stealing Africa*) or flout international conservation law (as in *Virunga*).

Stealing Africa (Christoffer Guldbrandson, 2012 – part of the Danish Why Poverty films project[42]) investigates why 60% of Zambia's citizens live on less than $1 a day even though the country is Africa's second largest copper producer and the world's third largest supplier of copper. Here is what the eco-doc uncovers. In 2000, the continuing fall in copper prices and lack of refurbishment of their mining assets meant that, in order to qualify for debt relief from the IMF, the Zambian government had to undertake privatisation. The outcome was that Zambia's share was reduced to a minority stake with no operational control and limited involvement on company boards.[43] In April 2000, Mopani Mines PLC purchased the assets. This joint venture comprises Glencore (an Anglo-Swiss multinational commodity trading and mining company, registered in St Hellier, Jersey) as the major stakeholder with 73.1%; the Canadian First Quantum Minerals Ltd with a further 16.9%; and the Zambian government with a paltry 10%. By 2011, Zambia copper mining had the largest reserves central to the global economy (in just ten years it amounted to $21 billion), but was owned by multinationals. For its part, government revenue on the registered company's profits represented a mere 18% of total tax revenues (equivalent to 3.5% of GDP) – a fact primarily attributable to the generous fiscal concessions given to the private mining companies at the time of privatisation, which included extremely low royalty rates, low taxes and generous capital allowances. The wealth goes everywhere except Zambia. One of those destinations, we learn right at the beginning of the film, is Switzerland. Ivan Glasenberg, CEO of Glencore (until December 2020, when he retired), presently a Swiss citizen, lives in Rüschlikon, a wealthy Swiss municipality where there is more tax revenue than expenses. This good fortune, we are told, is thanks to Glasenberg, who uses his tax receipts to fill the public coffers of his home town, thereby allowing the Mayor of Rüschlikon to lower taxes by 7%. This reduction is approved at a general Council meeting (which constitutes the opening [*ob*]scene of the film), despite a few citizens arguing that, surely, 2% should go to poor countries in Africa. *Stealing Africa*'s press release neatly summarises the eco-doc's findings:

> In this documentary, director Christoffer Guldbrandsen asks how it is possible for Zambia to make so little money from their copper when it is such a valuable commodity. He invites experts and politicians to shed light on the situation from various angles. Glencore is revealed to be a multinational with little name recognition that still controls 60 percent of the world's copper production. Meanwhile, the Zambians have begun to believe that the tax situation should be reformed. Guldbrandsen follows Guy Scott, the white vice-president of Zambia, who has made it his mission to ensure that proper taxes are paid for their copper. But with the slow judicial process required to change things, one should not hold one's breath.[44]

That 'little name recognition' is doubtless assisted by the fact that Glencore PLC, whilst being an Anglo-Swiss multinational company, has its headquarters in Baar, Switzerland (one international tax haven) and is registered in St Hellier, Jersey (yet another tax haven). For its part, Glencore's Mopani Mine is controlled through the British Virgin Islands (yet another tax haven). London is apparently the location of one of its subsidiaries. The meaning of all this will become clear in a moment. The background to this multinational (and its ghost qualities) begins, in the 1970s, with Mark Rich, a financier, commodities trader and hedge-fund manager who was indicted in the USA for tax evasion, fraudulent transactions, racketeering and much else (selling arms to Iran for oil, then a US enemy country, is but one example). It was Rich who set up 80 subsidiaries to his company, over 5 continents. This enabled the building of anything anywhere; the selling of one's own products to one of one's own subsidiaries for a pittance (thereby avoiding tax payments to the country of origin) and then selling on for huge profits. Thus, in this instance – despite Glencore PLC being an Anglo-Swiss corporation, but because it is registered in Jersey – London can act as a subsidiary (hence my term 'apparently' above, this is also true for a Swiss town). Currently, Glencore (under the over-arching company name of Mopani) sells its mined copper to its subsidiary-self in Switzerland (which has no use for it incidentally) at a cheap price (25% of its real value) and then sells it on for profit, all pretty much tax free.[45] These are known as transfer-pricing practices.

As vice-president Guy Scott points out, in *Stealing Africa*, Glencore's failure to pay their appropriate taxes means there is no investment in the country's economic infrastructure: no stable job creation, no social services, no medical services.[46] Zambia remains in debt (for example, at best it recoups 40% of its electricity costs for the mining process). Several attempts, in the early 2000s, to introduce new tax regimes, failed. The multinationals (Mopani) reacted by threatening to close the mines, or take Zambia to international courts. We learn also that, even when countries from outside scrutinise the figures and find they discriminate against Zambia, still the country is powerless to redress the imbalance.[47] For, the multinationals claim they need investment to improve their plants' efficiency and, therefore, make very little profit at all. Quite scandalously, in response to a 2004 independent investigation into the health and environmental effects of the sulphur-dioxide dust spewing into the local atmosphere and the acid leaching into rivers, Glencore applied to the European Investment Bank (EIB) for funding to build a new smelter, claiming it was an environmental necessity.[48] This $50 million loan, granted in 2005, was later repaid in 2011, but it is unclear to what ends it was put since continued environmental studies showed no improvement in air and water conditions. In 2009, the air pollution was 25 times above the WHO permissible sulphur-dioxide levels and the leaching into rivers (the local source of drinking water) of chemical waste was 16 times the permissible levels. The EIB funds are ostensibly reserved to help poor countries. In 2011, the Zambian Revenues Authorities' complaint to the EIB, of tax irregularities and fund misappropriation, was ruled against by the EIB Inspectorate General, a ruling endorsed by the OECD.[49] The message is clear. Poor countries cannot fight against transfer-pricing practices (a hugely complex procedure). Even if the evidence of fraudulent practices is there – as in the period 2008–11 when Glencore was successful in securing EU export refunds – the corporation is strong enough to dispute the claims of illegal practices. Moreover, investment banks and the

OECD have vested interests in protecting the status quo (as we saw with the IMF in relation to debt relief, cited above).[50] This self-serving attitude extends also to the indifferent behaviour of multinationals and government officials toward local communities – in Zambia's instance, the local inhabitants suffering health issues, and local farmers who had their land nearby the mining plants appropriated (with only a handful relocated).[51] At the time of the making of *Stealing Africa*, it was also the case that the then-President, Rupiah Banda, was keen to protect foreign investors and attract others (including China), and was very determined not to upset any existing apple-carts by demanding more tax revenue or better conditions for the local Zambians.[52]

And of course it's a similar story with *Virunga*: corrupt multinationals and corrupt officials, including presidents, this time in the DRC, willing, in the first case, to exploit economic precarity to get a grasp on the panoply of mineral and oil resources located in Virunga National Park and, in the second, to sell out on people and their environment for financial gain. The opening five minutes of the eco-doc make it perfectly clear what the roots to all this abuse are, colonialism and its destructive aftermath in the form of a post-colonial system that, on the one hand, reproduces the privileged practices of the former institutions and, on the other, witnesses a resurgence of tribalism which, up until this post-colonial moment, had been by and large controlled or repressed by the colonisers. I've already written about this issue extensively in *Film Ecology*, but suffice it to say that it is very important indeed to recall that most post-colonial countries have yet to step out of the shadow of their punitive past. A few exceptions aside (such as Botswana), African nations have fallen foul of the principles of democracy and the DRC is no exception. It has been a virtual dictatorship since Prime Minister Lumumba was assassinated in 1961 and Mobutu seized power through a military coup in 1965.

Since then, civil strife or war between rebel or dissident groups and the government has riven this country apart. A first Congo War in 1996–7 rid the country of Mobutu, only for an equally autocratic ruler Laurent-Désiré Kabila to take his place – with the help of Ugandan and Rwandan armed forces, eager to get their hands on North Kivu province (known for its rich resources, in particular cobalt, and home also to Virunga National Park). These armed forces briefly retreated to Goma on the border with Rwanda where they were joined by other rebel Congolese factions. Together they started the Second Congo War, 1998–2003, during which time Kabila was assassinated and his son Joseph Kabila replaced him (in 2001). Several accords were signed in late 2002. First, the Pretoria Accord whereby the Rwandan troops agreed to withdraw. The second, the Luanda Agreement whereby the Ugandan forces would withdraw. Implementation proved difficult, not least because rebel DRC troops were also holed up in the eastern region (North and South Kivu provinces especially) and so the Rwandan and Ugandan troops were enabled to flout the accords and slip back over the borders. Accusations that these troops were involved in pillaging and trafficking natural mineral resources, whilst denied, only served further to destabilise the DRC. By 2005 the United Nations International Court of Justice ruled that this illegal looting had taken place and called for compensation. Meanwhile the IMF and World Bank (remember these jolly fellows?) awarded debt relief packages to Rwanda and Uganda for improving their respective economies (overlooking the fact that this recovery was largely facilitated by the pillaged minerals from the DRC).[53] This looting is not limited to these insurgent troops. The DRC's own rebel troops, dissatisfied factions and mercenaries are equally after these assets to line the pockets of the rich both in the DRC and internationally.

In 2008, an armed rebellion against the government led by a former general of the DRC armed forces, Laurent Nkunda (who had set up his rebel troops in North Kivu), forced President Kabila to make a deal with the Rwandan government whereby Rwandan soldiers were allowed back into the DRC to put an end to the uprising and arrest Nkunda (in January 2009). This successful outcome led to a peace treaty being signed between the government and the rebel Congolese (known as the National Congress for the Defence of the Congolese People, CNDP). The CNDP agreed to become a political party and integrate its troops into the national army. However, by 2012, the CNDP claimed the government had not honoured their side of the treaty (the release of their imprisoned members) and formed a military rebel group M23 (March 23 Movement) and violence erupted once more – again in the North Kivu Province, the home to the Virunga National Park, a designated UNESCO World Heritage site.

It is precisely into this maelstrom of civil discord that Orlando Von Einsiedel arrived to make his documentary, *Virunga*. What started out, in early 2012, as a documentary about the Virunga park rangers risking their lives to protect endangered mountain gorillas and trying to rebuild their country through a number of development projects, very quickly turned into a much bigger story about illegal oil-prospecting by SOCO, a London-listed British oil company, poaching animals and charcoal, and the M23 rebellion. Here is what Von Einsiedel has to say (in 2014) about the genesis of his film:

> Between 2008 and 2012, there had been a period of relative stability in eastern Congo. I was looking for a story from the region that wasn't the usual war, misery, sexual violence – the kind of stuff that everyone reads about eastern Congo. I picked up a newspaper on a plane one day, and there was this story about the rangers of Virunga National Park rescuing a baby mountain gorilla. I thought, 'Wow, this sounds like a really interesting story' and the park was doing all these really ambitious development projects, and tourism was coming back. This story could act as a metaphor for the wider rebirth of eastern Congo as a whole. So, I packed my bags and went out there … what's special about the park is that it is people-focused. You can't really do conservation unless you address poverty. What really impressed me … was the way the park was trying to address that. It's doing

> lots of tourism and (run-of-river) hydroelectricity schemes. That's all about creating jobs that create stability and lasting peace.[54]
>
> *(my parenthesis)*

Once he arrived, however, this first idea ran up against another truth. As Von Einsiedel puts it, 'when I got there I learned very quickly of the park's concern about the oil company and I saw that what I could contribute to the fight was technology'.[55] And just four weeks after his arrival, a third element, the M23 rebellion, entered into the story's equation. The original human story about the park rangers' work and the promise this could deliver to their country dramatically morphed into something very different. To quote Von Einsiedel:

> In terms of the film, it started to look like three separate films: an investigative PBS 'Frontline'-type film, a National Geographic film and a kind of war movie. This suddenly became something that was so much bigger than anything I'd ever done before. But I always felt they were so interrelated they had to be told together.[56]

Indeed the stories are interrelated. In 2010, SOCO was granted permission by President Kabila to prospect for hydrocarbons in Block V in eastern Congo.[57] Much of that area encompasses Virunga National Park, a protected area, and Lake Edward (an important source of fish for the Congolese). SOCO parachuted in with promises, if the park was declassified, of schools and universities and other infrastructures, meaning roads – in short SOCO behaved much like the EHM we saw in *Dreamland*. But it met with a great deal of resistance from the park rangers, led by the chief warden of the park, Emmanuel de Mérode, and most of the local residents.

It took just over two years to shoot this eco-doc which rests, in terms of a structure, on a triangulation of different opposing forces, a process which Von Einsiedel describes as follows:

> I frame it in terms of how the park is affected. It has threats from poachers, it has threats from rebels. These are threats that can be managed and addressed. The oil threat is a threat to the entire integrity of the park, and if it continues it can destroy Virunga forever ... What SOCO is doing is bad on so many levels. It's potentially short-circuiting the possibility of Virunga ever reaching its potential. Then there's the concerns we raise in the film: illegally operating in a World Heritage site, their subcontractors and supporters are possibly involved in bribery and corruption, links to armed groups, and human-rights abuses.[58]

True to his commitment to contribute to the fight with camera technology, Von Einsiedel shot a great deal of footage during his first year there on a Canon 7D, a really inconspicuous camera. As the story advanced, he brought in the use of undercover cameras and with the help of French journalist,

Mélanie Gouby, and Rodrigue Katembo Mugaruka (the park's senior ranger) exposed the venal procedures SOCO employed to try and muscle its way into the park, bringing in mercenaries (under the guise of subcontractors) to do their dirty work of bribing officials, local military men and rebel forces. We learn also, through this footage, how SOCO, in its attempts to force its way in, was able to piggy-back onto the unrest caused by M23. In an undercover interview, the leader of the M23, Colonel Kazarama (a wealthy landowner) is heard to say he 'will make sure those against SOCO are kept down', adding that he stands to get a nice percentage of the oil prospecting. Against this background of violence, the rangers carry out their functions: protecting the wildlife, in particular the endangered mountain gorilla, often at the risk of their own lives (in the last 20 years over 170 rangers have been killed by rebels or poachers), offering sanctuary to orphaned gorillas (their mothers get killed by poachers so the babies can be stolen and sold on the illegal market). The humanity of the rangers is exemplified by the three central personages to this documentary narrative: Mérode, Rodrigue Katembo Mugaruka and André Bauman (the gorilla caretaker). All understand the risks they take (as do the other rangers) and Rodrigue Katembo Mugaruka sums up their ethos in these closing words to the documentary: 'I accept all that might happen to me, but what we wish for is for the park to live forever'.

Von Einsiedel talks of the transformative power of the Virunga Park which drew him to the project.[59] It is also the case that his eco-doc had its own transformative impact. 'We made this film', he states, 'and the one thing we thought we could do was gather evidence about the wrongdoing and publicise to the whole world what exactly is going on in the eastern Congo'.[60] The eco-doc was produced by Von Einsiedel's Grain Media in tandem with Joanna Natasegara's Violet Films – two small, social justice-committed companies. However, after premiering in numerous festivals in 2014, where it was richly awarded (and subsequently nominated for the 2015 Oscars), it was picked up by Netflix and shown to Leonardo Di Caprio who offered his support by brokering a distribution deal with Netflix (becoming the film's executive producer in effect). As Von Einsiedel explains, this brought in an audience 'who might otherwise not be interested in a film about the Congo, and it keeps the spotlight on these issues'.[61] The effect of this wide distribution was to make 'shockwaves in powerful circles'.[62] From Desmond Tutu to Hillary and Bill Clinton to SOCO shareholders Richard Branson and the Church of England, the outcry was significant enough to cause SOCO to stop its operations and withdraw – but not to rescind its contract, which it still retains to this day under its new name of Pharos Energy PLC (some lighthouse!). And sadly, as the civil unrest continues, so too Virunga Park continues to suffer. Since this film was released, a further 14 rangers have been killed by rebels, local militia or poachers; so too tourists have died. Poaching, looting and charcoal smuggling carry on. The recently elected president of the DRC, Félix Tshisekedi, shows no signs of cancelling oil-drilling contracts on the 'prospecting blocks' in the park (either with SOCO or the French company

Total). The park was briefly closed in 2018–19 but is now once more open to tourists, for how long one wonders.[63]

Virunga perfectly exemplifies, within a micro-context, the clash of two economic systems: on the one hand, extractive deregulated neoliberal capitalism, as embodied by the institutions of power (the president, the government, the army and the multinational SOCO); and, on the other, the regenerative circular economics system, structured to meet the needs of the local inhabitants, as proposed and implemented by the rangers' management policies (embedded as they are within society and within nature's carrying potential). This eco-doc provides us with a clear picture of the lengths to which the dominant economic model will go and proves, if needs be, how the Capitalocene is a fossil fuel economy that survives only by constantly expanding its tentacles, gobbling up the planet as it grows. This tenacious belief in the growth model seems unassailable. However, it reminds me of R. Buckminster Fuller's poignant words: 'you never change things by fighting the existing reality. To change something, build a new model that makes the existing model obsolete'.[64] And this brings me, once again, back to Raworth's Doughnut Economics model. Its beauty lies in the fact that it both reveals the harm our stubborn adherence to the current neoliberal economic system causes and renders self-evident, in so doing, the path we need to follow. A model that makes the existing model obsolete.

If we now transcribe what we have learnt, from the seven eco-docs discussed in this section, onto Raworth's Doughnut diagram, a very clear image appears (see Figure 1.1).

In terms of all four energy systems under scrutiny in this section on cheap energies (nuclear, oil, minerals, hydropower) and the nine planetary boundaries essential to the environmental health of the Earth, we note, first, that all seven eco-docs demonstrate how the exploitation of these energy sources has led to a transgression (or overshoot) of the ecological ceiling and contributed to climate change. Second, we observe the omnipresence, in all nine planetary boundaries (see direction of red arrows), of oil as the primary overshooter and despoiler of the environment (both in terms of production and its affiliated by-products). As for the nuclear, its impact on the ecological ceiling is in evidence in five boundaries: *freshwater withdrawals* (for cooling), seepage into *oceans*, *land conversion* (for building infrastructures and nuclear-waste burial purposes), *biodiversity loss* and *air pollution* (the effect of radiation, in both instances). Minerals exploitation impacts upon four boundaries: *biodiversity loss*, *air pollution*, *chemical pollution*, *freshwater withdrawals*. Finally, hydropower. Here there are two tales: the harm its exploitation can bring about in terms of *freshwater withdrawal*, *land conversion* and *biodiversity loss* as seen in *Undamaged* and with *Dreamland* the added problem of air and soil *pollution*. But there is also the positive outcome as indicated in *Virunga* where a small run-of-river hydro-electric power plant (opened in 2013) provides *clean energy* to communities living around the park – as denoted within the biosphere in the diagram.[65] A further point on *Virunga*, you will see that it is mentioned in relation to *biodiversity loss* and *land conversion*, and is placed between

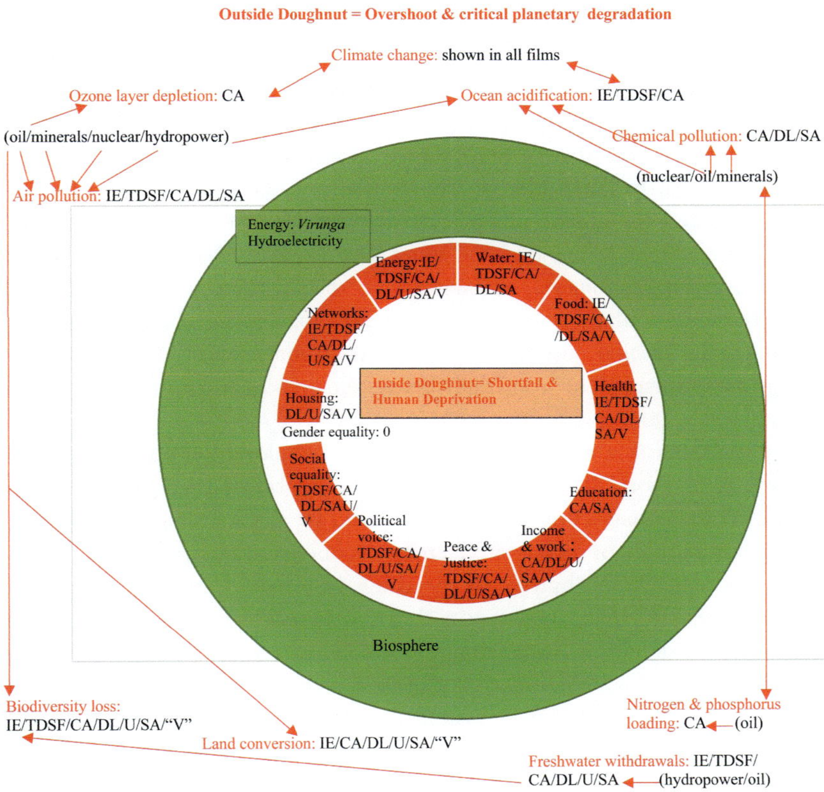

FIGURE 1.1 Mapping the seven 'cheap energies' eco-docs onto the Doughnut (IE = *Into Eternity*; TDSF = *The Day the Sun Fell*; CA = *A Crude Awakening*; DL = *Dreamland*; U = *Undamaged*; SA = *Stealing Africa*; V= *Virunga*).

quotation marks ("V"). I have done so because, whilst oil exploitation has not yet occurred in North Kivu province, nonetheless, the struggle for prospection, the actual mineral thefts and the civil war conditions in that area have already impacted upon these two boundaries. Most significantly, planting and harvesting are seriously disrupted by these continued aggressions with the ensuing poverty, starvation and internal displacement– one example of which we witness in the film when M23 capture Goma (the capital of North Kivu province) and 60,000 inhabitants seek refuge in Virunga Park.

And what of the impact of these planetary transgressions upon the social foundation? In 11 out of the 12 dimensions, there is a shortfall, leading to human deprivation (as indicated in red inside the Doughnut, in Figure 1.1. I found it rather remarkable (curious, even), however, that, in these seven eco-docs, the one dimension to get no mention of being impacted upon is *gender equality*. When we come to discuss *Solar Mamas*, we will see just how vital it is that this particular issue obtains address – after all, who is it who owns the energy resources? But, for now, let's stick with what we see revealed. Six of these eco-docs show how the pursuit of *energy* in all its manifestations impacts upon *peace and justice*, access to a *political voice* and *social equality* (*The Day the Sun Fell*, *A Crude Awakening*, *Dreamland*, *Undamaged*, *Stealing Africa*, *Virunga*). In this context, the most extreme effects of this pursuit at all costs to capitalise on resources are exposed in *Stealing Africa* and *Virunga*. In five films we are shown how the changes in *biodiversity*, *land conversion* and the abuses of *freshwater withdrawals* have impaired the quality of life in terms of *health*, *food* and *water* (*Into Eternity*, *The Day the Sun Fell*, *A Crude Awakening*, *Dreamland*, *Stealing Africa*); and, in the case of *Virunga*, with regard to these particular dimensions, we saw how local attacks on the park rangers endanger life (*health*), just as rebel forces' insurgence upon local inhabitants causes serious displacement from their ecosystems (and concomitant impact on *health* and *food*).

Energy itself, because it is a source of abuse and exploitation (of land as much as people), is in shortfall of the social foundation for all seven films – even though, as with *Virunga*, this abuse is countered by the successful hydropower plant which is a source of regenerative energy economy. The insecurity of *income and work* is touched upon by five of the films (*A Crude Awakening*, *Dreamland*, *Undamaged*, *Stealing Africa*, *Virunga*); a shortfall in *housing* in four (*Dreamland*, *Undamaged*, *Stealing Africa*, *Virunga*) and *education* in two (*A Crude Awakening*, *Stealing Africa*). Most significantly, six of the seven eco-docs point to the shortfall in the case of *health* and all seven where the *networks* dimension is concerned. *Health* is put at risk by the impact upon the environment of the energy practices of the nuclear, oil, minerals and hydropower industries. Practices, we observe, that are endorsed, enforced by those controlling or dominating the *networks*, namely large-scale corporations in the industrial sector, international banks or indeed government officials – bodies so big they squeeze out smaller players to say nothing of the lives of ordinary citizens. These powerful organisations, as we saw, were able to hide the truth about health risks, invoke the fear factor of job insecurity, behave in such a way as to undermine public accountability. When used to misinform, *networks* influence public opinion by disguising the truth and do not serve the interests of the biosphere (which is a safe and just space for humanity and the more-than-human world).[66] Moreover, such concentration of power in the hands of the few can only result, as Raworth points out, in a 'highly unequal and brittle economy' (176). A brittle economy that is stuck in the misguided belief of growth. An economy that ignores the fact that the ecosystem is not a machine, but sees land, instead, as capital whose resources are the ONE IDEA to be excavated.

But what if, instead – as both Kate Raworth the economist and Robin Wall Kimmerer, the environmental biologist, propose – we were to recognise the value of nature's networks; to learn the science of restoration ecology by emulating nature, by being in a reciprocal relationship with the Earth? Raworth speaks of the 'recurring success in nature's designs' whose 'networks are structured by branching fractals' that are 'just like tributaries in a river delta, branches in a tree', and whose 'resources such as energy and matter and information can flow through these networks in ways that achieve a fine balance between the system's *efficiency* and its *resilience*' (175, Raworth's stress).[67] Kimmerer uses very similar terms when she speaks of nature's design whose ecosystem is 'a community of sovereign beings' which begins with plants who are 'the oldest and most effective of land healers' (2013: 331). This is where we can learn reciprocal ecological restoration: 'with plants and natural processes in sole command, the role of land as a renewable source of knowledge and ecological insight becomes apparent' – their ability to adapt and to heal points to their 'ingenuity and wisdom' and thereby show us the way (333). In pursuing what Kimmerer terms a bio-regenerative economy and taking a new holistic approach which sees 'the land not as a machine but as a community of respected non-human persons to whom we humans have a responsibility' we would be engaging in what is known as a 'biocultural or reciprocal restoration' relationship with the planet (338). In brief, 'as we work to heal the earth, the earth heals us' (340). Indeed, we witnessed evidence of planet Earth's ability to heal herself during the brief world-wide lockdown period of March–June 2020 due to the COVID-19 pandemic – and, too, a greater awareness population-wide of the need for ecological responsibility. In a less dramatic context we shall see examples, in Chapter 3, of this reciprocal restoration practice recorded in the eco-docs relating to agroecology.

Let us now, however, turn to the ENERGY alternatives and the three eco-docs listed in column III of Table 1.1. The primary point they make is that we are all able, on however large or small a scale, to help the planet heal itself. The very title of Agnès Varda's film *Les Glaneurs et la glaneuse/The Gleaners and I* points to the fundamental truth of humankind's reciprocity with the land. Gleaning means a hands-on engagement with the earth's food system. But it also means to learn and, therefore, obtain wisdom, which in this film's instance, is about our relationship to waste. Indeed, Varda's eco-doc actively promotes the concept of 'no waste' through the principle of 'use-make-re-use' which we see advocated by Raworth's butterfly drawing of a circular regenerative economy (in Figure 1.2).[68] In this model, says Raworth, we consider all materials

> as belonging to one of two nutrient cycles: *biological* nutrients such as soil, plants and animals, and *technical* nutrients such as plastics, synthetics and metals. The two cycles become the butterfly's two wings, in which the materials are never 'used up' and thrown away but are used again and again and again through cycles of reuse and renewal.
>
> *(221)*

This is precisely what Varda's film explores in her travels across France as she encounters field gleaners, urban gleaners, artists gleaners (and others) – all recycling rejected objects into their own lives, as food, art, re-usable items; all seeing in what is left as waste its re-usable potential (in Chapter 2 I shall be discussing

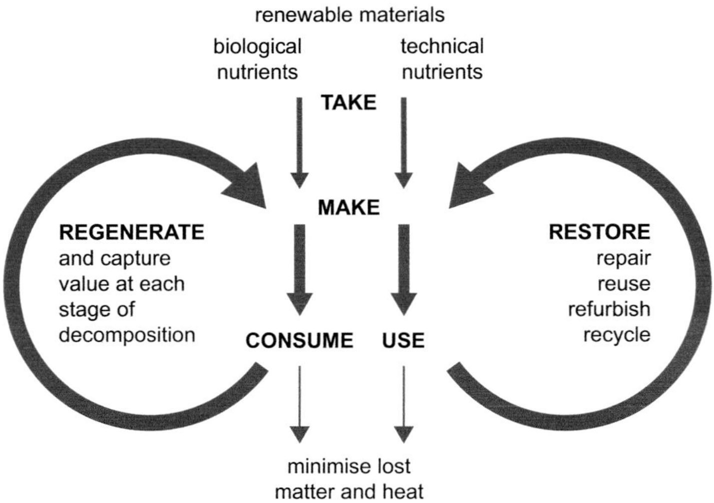

The butterfly economy: regenerative by design.

FIGURE 1.2 Raworth's butterfly diagram adapted from Ellen MacArthur's model in *Towards a Circular Economy*.

this issue in depth with *Waste Land*). This is what we can all engage with, at a small level, to bring change – reducing waste reduces energy needs. And here I must move on from Varda's film. Regrettably. But, for reasons of space and because the film has received a lot of critical attention, I've decided not to dwell further here on this illuminating film.[69] Instead, I have chosen to take on board the two eco-docs, *The 4th Revolution* and *Solar Mamas*, which very specifically address, within the domain of ENERGY, the reciprocal possibilities of a regenerative economy.

1.3 Regenerative economy – holistic energy practices and the restorative exchange-value of *The 4th Revolution* and *Solar Mamas*

Although *The 4th Revolution* (Carl A. Fetchner, 2010) and *Solar Mamas* (Jehane Noujaim & Mona Eldaief, 2012) approach the question of holistic energy practices from almost opposing ends (respectively, the macro and the micro), their main message is the same, namely, that energy autonomy (i.e., free from fossil fuels) can be derived from naturally occurring energy sources (such as solar, wind, geothermal) – sources which offer all of humanity equal opportunity. This idea of energy being harnessed to equality and democracy is indeed an exciting one, and provides us with the triangulated diagram in Figure 1.3.

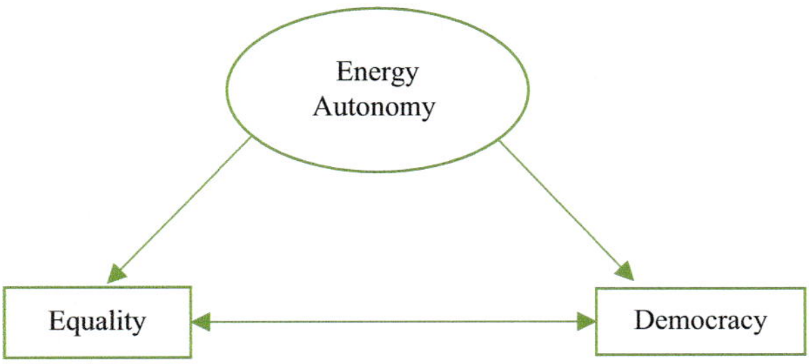

FIGURE 1.3 Triangulation of energy autonomy.

This concept is at the core of Carl A. Fetchner's *The 4th Revolution: Energy Autonomy*, almost as if echoing Buckminster Fuller's words to 'build a new model', and it sets out as its intention to explore what this triangulation means on a macro level. In the press-release notes, Fechner states that 'only a broad-based awareness of the possibilities of renewable energy can ignite an international movement and usher in the crucially necessary energy transition'.[70] For its part, *Solar Mamas* sets about with the same beliefs but on a micro level, showing how women – from impoverished countries and with virtually no education – can be trained to build solar panels and, thereby, bring all levels of social well-being to their small communities. The same ethos, the democratising of energy, is present in this eco-doc, but the challenges to its implementation, rather than being at the global level (of multinational corporations and international agencies' resistance as with *The 4th Revolution*), are, this time, at the local level of cultural constraints. However, more on this film in a moment. First, *The 4th Revolution*.

The 4th Revolution has as its starting point the work of the bio-economist, Hermann Scheer, whose words accompany the opening shots of Los Angeles' skyline of towering glass-clad buildings.[71] Scheer tells us that if all the glass-fronted office buildings had used, instead, solar panels they would have enough energy to run both building and office infrastructural needs and have a surplus which could be sold onto the grid. And this by and large is the overall message this eco-doc wants to deliver. We see several extremely successful examples of this practice in terms of solar, aeolian and geothermal energies being used to run entire communities and commercial sites. We also hear from engineers, some designing cars for the future and others a variety of possibilities for energy storage batteries.[72] We visit a reforestation programme in Brazil's Amazon forest, the Juma Sustainable Reserve Project, that aims to reduce the greenhouse gas emissions from the deforestation programme going on in the state of Amazonas.

Evidence builds upon evidence, this revolution *can* take place – is already on the march, despite what the oil and coal lobbies say through their mouthpiece, the powerful International Energy Agency headed up by chief economist, oil and energy analyst and CEO Fatih Birol –interviewed in this film – who considers that a transition to 100% renewables is unrealistic and sees the future in terms of (and I quote) 'clean coal' energy and 'secure nuclear power'.

Let's take a closer look. Fetchner travels the continents with his small five-person crew meeting people of diverse backgrounds all committed to the democratisation of the energy supply (be they politicians, activists, financial entrepreneurs, scientists, mid-wives) and gathering footage, what he calls a 'visualising sustainable activity', which gives proof that 'the transformation of the world's energy supply to 100% renewable energy is possible' – footage which includes in-depth solutions to the 'most challenging problem of our time'.[73] Amongst the many extraordinary feats of inventiveness, I will discuss just four: Preben Maagaard's aeolian Nordic Folkcentre for Renewable Energy in Denmark; Matthias Willenbacher, co-founder in 1996 with Fred Jung of the juwi Group in Germany, a leading entrepreneur in regenerative energy supply; the financier Muhammad Yunus, founder of the Grameen Bank, in 1983, which provides micro-loans to the rural population of Bangladesh (95% of whom are women[74]); and, in a brief, nuanced way, Elon Musk, a leader in electric car technology.

Maagaard and Willenbacher demonstrate through their installations how 100% renewable energy works. As early as 1983, the Danish renewable energy pioneer Preben Maagaard had shown that energy autonomy was do-able – his wind-energy research was driven in direct response to the various 1970s oil crises. Presently, 50,000 people draw their energy from 100% wind energy generated by this Folkcentre. The centre is also a training centre which, since 1986, has formed generations of renewable energy experts all over the world. One such trainee, Ibrahim Togola from Mali, explains how this programme inspired him to install autonomous energy systems for the rural populations in Mali, allowing them to access electricity which, in turn, opened up economic possibilities and of course enabled children to pursue their studies at home in the evenings. As Togola remarks 'our country has so much potential, but nothing works; 80% of the population is without access to electricity, this is the missing catalyst; with it we have a chance' – making clear the link between poverty and lack of autonomous energy resources in what are currently termed 'developing' countries (formerly known more explicitly as Third World economies).

As to Willenbacher, the best example for the visibility of his vision is his company's juwi Group's construction of the world's most energy-efficient building in Wörrstadt. As we are escorted around the building, Willenbacher explains how its architecture, fixtures and fittings all comply with the company's vision of conserving resources. The environmental impact of the whole building is exemplary: heating, water, electricity and water-aided air-conditioning are entirely driven by regenerative sources (solar, wind and bioenergy). Furthermore, the energy efficiency of this type of construction is such that every element used means that the building pays for itself within five years.[75] Figure 1.4 provides a picture of this beautiful building.

Amusingly, Willenbach proudly steers his Tesla car to its electric carport as we segue into an exploration of his building after an interview with Elon Musk,

FIGURE 1.4 juwi Group headquarters, the world's most energy-efficient building, Wörrstadt, Germany.

Tesla's product-architect, in which he explains that his Tesla model is but the prototype for an affordable and stylish electric car. The future is the total concept of electrically recharged cars and homes, he declares, all driven by renewable energies which would liberate us from the oil industry (of course, given Musk's other SpaceX-driven ambitions, launching rockets to the moon and Mars for example, he sadly does not stand out as an ecologist, more as an eccentric adventurer).[76] His prognosis of electric vehicles (EV) as a viable transportation system for the future, however, seems a judicious one. Since the date of this film, 2010, the EV market has expanded from $100,000 to $5.6 million in just ten years.[77]

At the other end of these macro-renewable energies projects, *The 4th Revolution* presents us with a number of micro-level projects working to the same end of energy autonomy and fiscal independence – a primary example of which is Nobel Peace Prize winner Muhammad Yunus' financial enterprise, the Grameen Bank (Grameen meaning 'village') – a micro-finance organisation and community development bank which makes small loans to the poor without requiring collateral. The banking system, we are told, is built on mutual trust, accountability, participation and creativity. Indeed, Yunus' starting point is that all human beings have creative energy and we only remain poor if we cannot unleash that energy. 'Poverty' he states 'is not created by poor people' and, given an opportunity, such as micro-credit, this 'helps to set them out and pull on their creative energy'. The success has been tremendous, particularly amongst the Bangladeshi women who have proved to be astute entrepreneurs and reliable borrowers. As Yunus says, 'because they are the householder, this means the investment goes into the home'. The micro-funding has enabled them not only to raise their status, it has lessened their dependency on their husbands, and through hard work helped to improve their homes and the nutritional and general health standards within their families. Recently, the bank funded training for women in solar energy (making solar panels for their homes) – and we see a most heartening outcome, a woman installing a panel for her home. The venture – originally referred to as 'a small bubble of hope' – is a remarkable story demonstrating how, even at a micro-level, we can meet the 12 elements set out on the social boundary of Raworth's Doughnut. Even though in 2011, a year after this eco-doc's release, Yunus was forced, in a politically motivated campaign, to resign, micro-financing continues to bear results. Grameen was a catalyst for micro-credit expansion and, since the late 1990s, some 3000 organisations have joined similar schemes (in India, Bangladesh and the Philippines, primarily). Findings from micro-credit studies, in 2019, indicate that, although micro-financing is not as impactful as in its early years, nonetheless, it still has an important role to play for the poor with modestly positive results.[78]

Weighing in against this series of positive examples of regenerative energy practices is the rhetoric of the fossil fuel conglomerates, with at its head Fatih Birol who very adamantly *denies* that our present reliance on fossil fuels is an existential problem (in that they are running out). The conglomerates, with OPEC safeguarding their pricing strategies, remain powerful opponents to change and exercise their lobbying rights over politicians with little regard to the environment's future.[79] Here's just one example provided by *The 4th Revolution* which shows why resistance is so strong. Fetchner explains that '2bn litres of fuel are sold worldwide by the oil industry each year for transport and traffic; that's one quarter of the world's energy consumption', so it is not only the fossil-fuel magnates who are disinclined to explore other system but car industries – as long as there's oil, why bother. This contemptuous attitude is of course driven by short-termism (OPEC is well-aware, as I mentioned earlier, that oil is running out). It's all about making profit now – revenues from oil in 2008, the time of this eco-doc, were $1 trillion – with Exxon a big winner (at $45 billion). So where is the incentive for these oil companies, who are inextricably linked to the current car industry, to promote electric cars? Indeed, once electric power comes from renewable energies – that would be the death knoll for one of the oil industry's major profit sources.

According to Fatih Birol, coal is in an equally strong position – and, uncannily, he correctly predicts that between 2008 and 2020 there would be a significant rise in coal power plants. Pointing to China and its energetic needs to fuel its economy, Birol states that the expansion is 'inevitable' and adds that 'coal will be the backbone of energy'.[80] Presently, it is a similar story in Germany and Poland. When Birol is questioned how to make coal less damaging, he glibly answers 'by carbon capture and storage'. Not only do the 'how' and 'where' and (of course) the 'who' constitute an almost impossible set of questions to answer, just imagine the costs involved! As a slight concession to the issue of CO_2 emissions, Birol proposes that 'nuclear power is a solution', and that there 'are no problems in terms of security' – we know how untrue that is, as *The Day the Sun Fell* made clear. However, when he remarks that 'people won't be pushed by moral arguments, but financial incentives', he unintentionally (to my mind) gives the key to what regenerative economics must unlock, namely, our willingness to change. To that effect, just recently, Fatih Birol seems to have made a U-turn, in his rhetoric at least. In 2021, eleven years after *The 4th Revolution*, he is now saying this rise in CO_2 emissions 'is shocking and very disturbing. On the one hand, governments today are saying climate change is their priority. But on the other hand, we are seeing the second biggest emissions rise in history. It is really disappointing'.[81] Despite this rise, it appears that the year 2020 witnessed a step change for the renewable energy industry which brought Birol to declare that governments need to 'build on this promising momentum' by putting in place policies that 'encourage greater investment in solar and wind, in the additional grid infrastructure they will require, and in other key renewable technologies such as hydropower, bioenergy and geothermal'.[82]

Solar Mamas (part of the Danish Why Poverty films project) is a documentary telling the true story of Rafea, an illiterate Jordanian Bedouin woman, living in poverty, who was selected, along with another woman villager (Umm Badr, her aunt), to take part in a six-month training course for the construction and installation of solar panels. The scheme, which is influenced by the Ghandian philosophy of each village being self-reliant, is run by Bunker Roy in India. In 1972, he set up this voluntary organisation, known as Barefoot College, which is dedicated to developing skills, promoting women's empowerment, providing training in education, health and electrification through solar power for the upliftment of rural people.

The course requires Rafea to commit to the six-month residency in India, and while she is desperate to attend – so that she can achieve both her desire to help her village (by installing panels and training other women to assemble and install the panels) and her dream of affording a house and providing a better life for her four daughters – she is met with the stonewall refusal of her husband, Alian, to let her go. Under Bedouin law, she must obtain his permission. Finally, pressure from the village Elders (one of whom is her enlightened father) persuades him to let her go. Also under Bedouin law, she and her aunt must be accompanied by a male member of the family. Umm Badr's son, Badr, is selected (which, incidentally, means he too can take the course, more on this in a moment). Rafea leaves her children in her mother's care; Alian refuses to help – asserting that, in accordance with Bedouin culture, his responsibility is to his first wife and not Rafea (his second wife). Since he has no work, this does not mean a great deal. However, again in terms of Bedouin law, as the father of Rafea's children, there is a great deal to concern her – for he can at any moment take her children away from her. This threat hangs over Rafea as she takes her leave of her village, with her disgruntled husband looking on, aloof from the villagers waving her off.

Once in India, Rafea blossoms, she is a quick learner, she befriends the other women (there are 27 women in all on the course, from Kenya, Burkina Faso, Congo, Guatemala and Columbia). We see them socialising, singing and dancing, as well as supporting each other in the tuition classes. Rafea no longer wears her headscarf at all times, perhaps because many of the other women don't – but perhaps also as an emblematic sign of the sense of freedom she experiences. Meantime, Badr proves to be a very poor scholar, reticent to put in any effort, despite the other women's willingness to help him. His indolence sharply contrasts with the women's commitment and most immediately mirrors that of his uncle Alian back home.

We soon gain a sense of the emotional cost of Rafea's decision to commit to the course, however. She learns over cell-phone conversations with her mother that Alian never comes to see their girls but leaves all their care up to her mother. And a month into the course, Rafea is forced to return home. One of her daughters is very ill, her mother informs her. A distraught Rafea prepares to depart, surrounded by her college women friends who try to console her, only to discover, once back in her village, that she has been lied to. Alian had pressured her mother into the lie by threatening to divorce Rafea and take her children away if she did not return.

But Rafea is now considerably more empowered than when she left and begins to speak her mind. First, she visits the Ministry of the Environment Facilitator, Raouf Dabbas – who selected her in the first place, and is distressed to see this golden opportunity come to nothing. He promises to intercede on her behalf so that she might return. Second, she explains to the villagers the value of solar energy to the betterment of all their lives and how she will be able to train them in these skills. Fortunately, at least the villager's chief Elder, Abu Badr, grasps the importance of what she says. Third, she declares to her mother: 'with or without your support, I am going back, I want to work'. Finally, she announces to Alian 'do what you want, I'm going' – thereby calling his bluff. As she says, 'by pretending I didn't care if he took the kids, to take everything, I deprived him of any control he had over me'. In the end, Raouf Dabbas (in a face-saving exercise for her husband) manages to persuade Alian to let her continue her studies and she heads back to India much to the huge delight of her women solar-mamas companions who quickly bring her up to speed on what she has missed at the college.

All the women successfully complete the course. Only Badr fails to pass. Rafea comes back to her community, her hair uncovered. Perhaps her joy both at succeeding and in coming first in the exam caused her to forget, but she is quickly reminded to cover herself. Undaunted, she shares her happiness with the assembled villagers: 'before I left I was an uneducated woman who knew nothing; with this knowledge I am going to change the role of women in Manshiat Al-Ghayath!' And indeed, in the closing scenes we see her talking to other women ('men and women are equal these days, tell your husbands you want to learn'); receiving the solar panel equipment from Barefoot College; assembling and installing a solar panel on the roof of her newly built home. We learn from the coda to the film that 'Rafea continues to pursue the solar project (in one week she assembled and installed some 80 panels along with Umm Badr) despite ongoing obstacles from the men in the village'. The nature of these obstacles is made only too clear in the brief meeting held by Raouf Dabbas with the villagers upon Rafea's return. Dabbas declares an association is necessary for things to move forward; at which point Badr immediately proposes himself as the President of the association (something he had already been asserting when in India at the college where he so singularly failed to pass the exam). Instead, it is the village's chief Elder Abu Badr (who has supported the scheme all along) who is appointed. Umm Badr is made Manager, and Rafea Field Co-ordinator of Training, responsible for spreading the knowledge. In an attempt at inclusivity, Dabbas makes Badr and Alian members of the association – a privileged position both men later abuse.[83] Against these difficulties, we are told, Rafea has managed, nonetheless, to continue with her project.

Solar Mamas is constructed around three essential elements, providing an additional triangulation to the one we saw above in relation to the democratisation of energy (see Figure 1.3). The starting point, this time, concerns the processes of empowerment and the cultural contexts with which these processes must negotiate before any democratisation of energy can be achieved. This is shown in Figure 1.5.

The red arrows in this diagram serve to indicate the pressure emanating from the cultural contexts of the Bedouin customs which are primarily coded within a strongly hierarchical patriarchal tradition. Gender inequality begins formally at the age of ten for a girl who is no longer permitted to attend school. As Rafea informs us, 'it is considered shameful'. Thus, women are contained within a state of illiteracy. As for the traditions surrounding family and marriage, as with all wives, Rafea has to obtain the consent of her husband to proceed with any plans; she must also seek the approbation of her parents with any undertakings that operate outside of her role as mother and wife; she must be accompanied by a male member of the family if she travels outside of the village. In these instances, patriarchy contains women within a subaltern status since they clearly have no agency and very few rights within Bedouin cultural hegemony. Considering that all 300 people living in Rafea's village are unemployed, it seems quite astonishing that this opportunity for her and her aunt to gain skills that would enrich the village meets with such resistance when it is first mooted – a point Rafea herself makes. We see, in the early scene, in which Bunker Roy and Raouf Dabbas try to explain to the villagers the advantages of the Barefoot College Training Scheme in Solar Technology, just how far this resistance goes. It is not just the menfolk (with the two exceptions of Abu Badr and Rafea's father), but also their

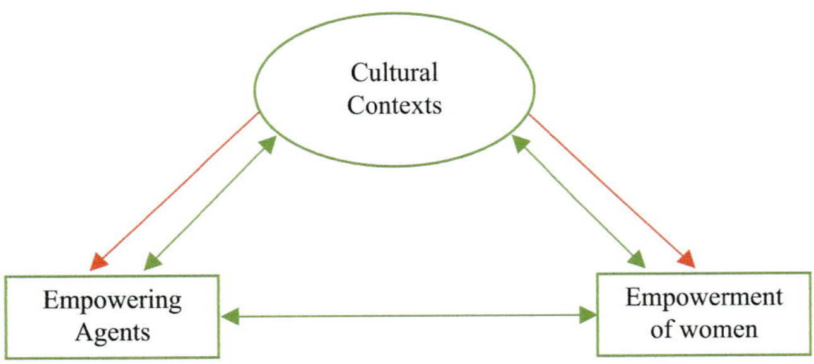

FIGURE 1.5 Triangulation of the processes of empowerment and cultural contexts within *Solar Mamas*.

wives who endorse the patriarchal rulings that such a departure is impossible. Even when Rafea is brought back, she has a hard time convincing the villagers how much they can benefit from the training she is receiving. Her words to her mother, 'I want to explore the world, to learn, to know how other people think', fall on unimpressed ears. For her mother, a wife's place is at home with her kids, adding that 'living like this is enough'. But, by now, Rafea's sense of empowerment (after just one month at the college) gives her the strength to reply to this staunch conservative position of no change. As she points out 'without education or purpose in life' and living in poverty, this is no life for her children. And, when she then declares she is going back regardless, this becomes her first major step towards independence and autonomy.

The green double-edged arrows in the triangulation of empowerment (Figure 1.5) indicate this powerful sense of agency Rafea experiences – one that has been supported by Dabbas' great faith in her ability. The culmination of this experience of empowerment comes not just with her successful completion of the course, but also with her refusal to allow Alian to 'kill her spirit' any longer through his determination to control her. She tells him quite clearly 'I haven't interfered in your life, but you interfere in mine ... you can't tell me what to do, I shall continue the project I started'. In much the same tone, Umm Badr tells the village's menfolk, 'you can't make us suffer'. The empowering agencies – Roy Bunker's Barefoot College, Raouf Dabbas from the Environmental Ministry and Abu Badr, the village Elder – have stood firm alongside these women, stepping in (certainly not taking control[84]) when necessary to facilitate acceptance, able therefore to respond both to the cultural contexts *and* the desire to empower these female solar engineers. This balancing act is reflected in the temporal structure of the documentary which devotes equal time to the two central elements making up this exposé on energy autonomy – the empowerment ethos of the scheme and the struggle for gender equality in a culturally conservative patriarchal environment.

Harnessing free energy is at the heart of these two very distinct eco-docs. With *Solar Mamas* the major outcomes are on a micro, local level, but they have huge social implications for the self-sufficiency of the villagers: economic independence for the women solar engineers, a community-based system of payment, clean and constant energy for all, light in the hours of darkness allowing children to continue with their studies, a training programme for women serving to stimulate their role in the local economy and reduce poverty. With *The 4th Revolution* energy autonomy is on a macro, planetary scale, presenting, as it does, the case for a 100% dependency on renewable energies within the next 10 to 20 years. Solar, wind, wave, biomass, geothermal, run-of-river hydropower – these are ready sources to be harnessed making it possible to construct zero-energy buildings, net-zero-water factories and passive housing (all examples of which we see in this eco-doc). Taken together, these two eco-docs reveal how respect of the ecological ceiling and the use of renewable sources can benefit all 12 elements of the social foundation, as you can see in the Doughnut diagram in Figure 1.6.

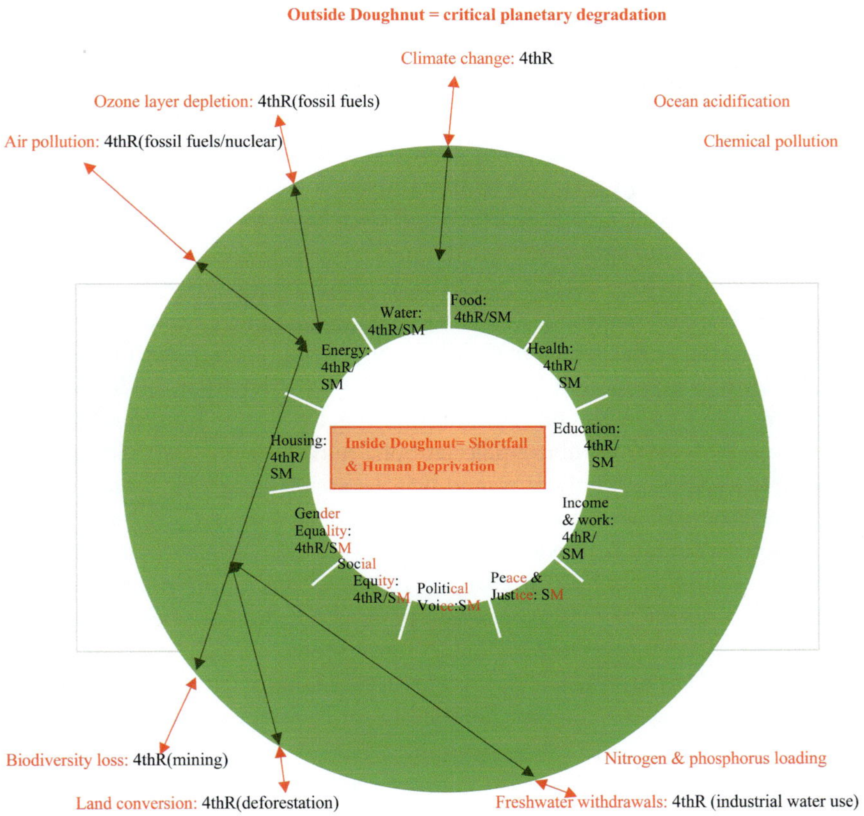

FIGURE 1.6 Mapping *The 4th Revolution* (4thR) and *Solar Mamas* (SM) onto the Doughnut.

The visualisation of the eco-docs' narratives illustrates just how the biosphere can be maintained, as a safe and just space for all, thanks to a regenerative and distributive economy based upon energy derived from renewable sources – this process is indicated by the green double arrows in the Doughnut spreading out from the energy box towards the perimeter of the planetary boundary, so as to represent the responses made to counter the dangers of planetary boundary transgressions. *The 4th Revolution* presents us with a more complex image than *Solar Mamas* since it directly addresses the forces that put the planetary ceiling and, therefore, the biosphere at risk – these are indicated by the red double arrows outside the Doughnut, showing the harmful energetic practices that threaten the Earth's well-being. In this eco-doc, the bio-economist Hermann Scheer unhesitatingly asserts climate change and the energy crisis are rendered much larger and expensive to combat because current political, technical and economic institutions are presently unwilling to embrace the realistic alternative of renewable energy. However, as this film and *Solar Mamas* make plain, the place to start mobilising this change of thinking is by giving evidence of effective renewable energy practices and by sharing this knowledge world-wide, rather than privatising knowledge for profit, as most countries within the existing Western neoliberal capitalist model currently do. In *The 4th Revolution* and *Solar Mamas*, people travel far-and-wide to see renewable energy at work and to discover how to harness this knowledge for their own country, town or village's well-being: Malian engineers come to Denmark and to Bangladesh to witness wind, biomass and solar systems producing energy; uneducated women come from all continents to India to train as solar engineers; Germans travel to Spain to see a model solar plant (200 hectares) supplying electricity to 45,000 homes – a model that could easily be adapted to the deserts in California for example (to say nothing of Saudi Arabia, which is considering this possibility). Using water from condensation or circulating used water to cool buildings, producing biogas from food surplus, developing biofuels from non-edible Jatropha nuts, refitting existing buildings with energy-saving technology, providing solar architecture new-builds (known as passive housing), developing battery technology, these are just some of the ways *The 4th Revolution* answers the voices of the 'disbelievers' or 'unshiftables' so readily embodied, at least in 2010, by the CEO of the International Energy Agency, Fatih Birol (as we have already pointed out, he has since changed his view on the value of renewables).[85]

As for *Solar Mamas*, we can determine from the Doughnut diagram (Figure 1.6) just how significant the training programme in India turned out to be for the women who attended the course. Certainly, Rafea's story is the dominant one and we know from what we witness that, as far as her well-being is concerned, all 12 dimensions of the social foundation are met – nearly all of them fully. Where there is red-letter marking on the diagram, this indicates there is still some shortfall. Yes, in terms of *gender equality, social equity*, having a *political voice* and obtaining *peace and justice* for herself and her girls, Rafea does achieve a degree of economic independence and assertion of her rights as a woman. However, it is clear from the closing scenes of the film that these equalities and rights are far from fully acquired – not just for herself but for the other women and girls in the village. It is therefore but a beginning – as Rafea herself declares: 'I shall continue the project I started; tell your husbands you want to learn; what is life without an education'. She is the first female solar engineer in Jordan – and in 2014 she was elected as the first female member of her municipal council in Manshiat Al-Ghayath. Thus, when she asserted, in the film, upon her return to the village: 'I am going to change the role of women in Manshiat Al-Ghayath' she was correct. *Solar Mamas*, just as much as *The 4th Revolution*, proves that all paths to protecting the biosphere have their impact. The link between poverty and energy shortage is a proven fact; energy access and empowerment through education is also a proven fact. The value of harnessing free energy is autonomy for all and a democratisation of the energy supply. Such a transition to decentralised and sustainable energy can only bring in its wake shifts in power, however gradual these may be. A shift that will progressively expose the constraining nature of capitalism's current model of politics-technology-economics which is so heavily attached to GDP/GROWTH, and, in place of this ONE IDEA-myopia, a will to change to new models, based in regenerative economics, must surely gain purchase.

Notes

1. See: <https://www.theguardian.com/environment/2021/feb/09/fossil-fuels-pollution-deaths-research> (accessed 9 February 2021).
2. I refer the reader to Regenia Gagnier's *The Insatiability of Human Wants* (2000) – in adapting her title to my own text I acknowledge the value of this study of late nineteenth-century economics and aesthetics in its tracing this transitional moment in Western thought from models of production to models of consumption.
3. Once, rainforests covered 14% of the Earth's land surface, now they cover a mere 6%; 60% of deforestation is for agriculture, 20–25% for commercial logging and the remaining 15–20% for mining. See: <https://rain-tree.com/facts.htm> and <https://wwf.panda.org/knowledge_hub/where_we_work/amazon/amazon_threats/other_threats/amazon_mining/> (both accessed 23 July 2020). *Home* informs us that land conversion for large-scale agricultural production accounts for 80% of all forest loss; that 90% of their crop yield goes to Europe or China; 80% of humanity's water consumption goes on agriculture; 80% of the world's mineral wealth is consumed by 20% of the world's population; 75% of the oceans' fish fields have been wiped out; one in ten rivers world-wide no longer flows all year.
4. Wetlands currently cover 6% of Earth's land surface, but according to reports 35% of them are disappearing world-wide (<https://thinkglobalgreen.org/wetlands/> accessed 23 July 2020). Wetlands contain one-fifth of all the carbon on Earth, about the same amount of carbon as presently exists in the atmosphere – a clear indication as to how careful we need to be with all these carbon time-bombs we are releasing into the atmosphere through our drilling into the Earth's crust, deforesting rain forests, cutting down trees (and so on and on).
5. This system allows for long-lens shots with rock-solid stability. The provenance of this technology is, unsurprisingly, military. As I indicated in *Film Ecology*, the interrelationship between war technology and the film industry is nothing new (Hayward, 2020: 6). The cameras were manufactured for army firing equipment and the gyro-stabilisation system for mounting the cameras (for example, on tanks or destroyers) to reduce vibration. The camera is integrated in the carbon fibre housing that rotates a full 360 degrees, allowing for full surveillance. For an idea of how this looks and functions see: <https://www.infinitioptics.com/technology/gyro-stabilization or https://gyrostabilizedsystems.com/broadcast-series or https://xd-motion.com/gyro-stabilized-heads/gss-cineflex/> (accessed 21 July 2020).
6. To hear the entire score see: <https://www.youtube.com/watch?v=oqczuvQy15U>. And for details on the full score list see: <https://www.cinezik.org/critiques/affcritique.php?titre=home_docu> (both accessed 24 July 2020).
7. Stylisation and excess are hallmarks of Besson's own film work, see his *Le Grand bleu* (1988) and *Atlantis* (1991), to say nothing of his signature-opening fast tracking shots across Parisian streets (*Nikita*, 1990) or into New York (*Léon*, 1994).
8. The film's release date was 5 June 2009 (World Environment Day), downloadable on-line, given free to TV and cinema theatres. <https://en.wikipedia.org/wiki/Home_(2009_film)> (accessed 24 July 2020).
9. Again a tone that sits uncomfortably with the funding context.
10. For further details on these measures, see: <https://en.wikipedia.org/wiki/Kering> (accessed 26 July 2020).

11 If you think about this for a moment it is a crazy example of how money becomes a commodity in and of itself. So, in the case of privatisation: the government sells *our* common (public) wealth; public wealth becomes private revenue for the rich elites to exploit as they see fit. In selling, the government also 'offers' the public limited shares in the privatisation, so we buy part of our original wealth back (thus doubling our original loss); we then receive a tiny percentage of profit on our original common wealth. And would it interest you to know that 'at least $18.5 trillion is hidden by wealthy elites in tax havens worldwide representing an annual loss of $156 billion tax revenue, a sum that would end extreme poverty twice over' (Raworth, 2017: 277)?

12 To see how these various systems work see: <https://en.wikipedia.org/wiki/Hydropower accessed 03/08/2020> (accessed 26 July 2020). It provides handy illustrations of all hydropowered systems.

13 This has yet to come to be. Until it was banned in 1993 most nuclear waste-producing nations dumped a considerable percentage into the ocean in waste containers and there have been incidences of leakage. See further: <https://en.wikipedia.org/wiki/Ocean_disposal_of_radioactive_waste> (accessed 3 July 2020).

14 Fairly alarming therefore wouldn't you say? See: <https://www.scientificamerican.com/article/what-does-the-us-do-with-nuclear-waste/> (accessed 3 July 2020). Equally alarming is the fact that, thanks to global warming and ice-cap melting, two of the US military's nuclear site projects of the 1960s, thought to be buried forever (!) under the Arctic ice-cap area (Project Chariot, to test the effect on water sources of radioactive contamination of tundra plants which is the major food resource for caribou… so right into the food chain; and Project Iceworm, for feasibility as a missile launch site) are now being uncovered and their nuclear waste has already (since the 1990s) leached into the environment; in particular in the case of Project Chariot, causing tundra sickness amongst Iñupiaq communities. See: <https://www.jstor.org/stable/40316687?seq=1> and <https://www.theguardian.com/world/2016/sep/27/receding-icecap-top-secret-us-nuclear-project-greenland-camp-century-project-iceworm> (both accessed 9 May 2021).

15 For more details see: <https://www.scmp.com/news/asia/east-asia/article/1848514/now-big-question-where-will-japan-dispose-its-nuclear-waste> (accessed 3 July 2020).

16 See: <https://ypte.org.uk/factsheets/sea-pollution/radioactive-waste> for more detail (accessed 3 July 2020).

17 <http://www.narmada.org/gcg/gcg.html> (accessed 26 July 2020). Sadly, dams as weapons of destruction are not limited to this single case. Recently, in Brazil (another BRIC economy) there have been two dam disasters, the Mariana Dam in 2015 (killing 19 people) and the Brumadinho dam in 2019 (with 248 deaths) – both as a result of weak regulatory structures. Moreover, metals in the tailings of these collapsed dams will be absorbed into the rivers' soil and affect the whole local ecosystem.

18 See: <https://en.wikipedia.org/wiki/Peak_uranium> and for general information on uranium see: <https://www.world-nuclear.org/information-library/nuclear-fuel-cycle/uranium-resources/supply-of-uranium.aspx> (both accessed 4 August 2020).

19 Conflicts date back at least to the Oil Campaign of World War Two with both the Allied and the Axis countries endeavouring to secure access to oil. More recently we can cite the Gulf War (1990–1) and the Iraq War (2003).

20 See Carlo Koos & Matthias Basedau (2013) 'Does Uranium Mining Increase Civil Conflict Risk? Evidence from a Spatiotemporal Analysis of Africa from 1960 to 2008' *Civil Wars*, 15:3, 306–31, DOI: 10.1080/13698249.2013.842744. <https://www.researchgate.net/publication/256034119_Does_Uranium_Mining_Increase_Civil_Conflict_Risk_Evidence_from_a_Spatiotemporal_Analysis_of_Africa_from_1945_to_2010> (accessed 4 August 2020). But, of course, uranium has been used in warfare. For example, during the two Iraq wars, depleted uranium was used in the so-called 'Dirty Bombs' by the British and American Forces. See: <https://ww

w.wagingpeace.org/the-real-dirty-bombs-depleted-uranium/> (accessed 4 August 2020).

21 See: <https://en.wikipedia.org/wiki/Nuclear_power_in_Japan%3e%20%20%20%20(4%20August%202020)> (accessed 4 August 2020). Presently, Japan has 39 operable reactors of which 9 are currently in operation.

22 See David Ratcliffe's illustrated interview of Michael Madsen's film *Into Eternity*, in: <https://ratical.org/radiation/IntoEternity/> (accessed 11 June 2020).

23 Some of which appears in Alain Resnais' film *Hiroshima mon amour*, 1959. Incidentally, American footage (in colour) of the impact and the radiation effects was not released until the 1980s.

24 This whole unsatisfactory response is fully detailed in <https://en.wikipedia.org/wiki/Fukushima_Daiichi_nuclear_disaster>. See also: <https://journals.sagepub.com/doi/pdf/10.1177/0096340212440359> (both accessed 20 August 2020).

25 See: Amory Lovins (2011) 'Soft Energy Paths for the 21st Century', <https://web.archive.org/web/20131224094312/http://www.rmi.org/Knowledge-Center/Library/2011-09_GaikoSoftEnergyPaths> (accessed 5 August 2020).

26 See: <https://www.armscontrol.org/act/2018-10/features/japan's-misguided-plutonium-policy and https://allthingsnuclear.org/gkulacki/does-japan-have-nuclear-weapons> (accessed 26 August 2020).

27 See: <https://foreignpolicy.com/2019/09/19/foreign-investors-fueled-violence-corrruption-south-sudan-africa-china-business-salva-kiir-arms-sales-oil/> (accessed 5 August 2020).

28 In the case of Iraq it is estimated at 4.6 million refugees. <https://www.pewresearch.org/global/2016/10/18/conflicts-in-syria-iraq-and-yemen-lead-to-millions-of-displaced-migrants-in-the-middle-east-since-2005/> (accessed 5 August 2020).

29 A more recent case of oil spill, this time with Shell in the limelight, causing damage by oil-extraction in the Niger delta, has been brought to the English Supreme Court. A move we might hope towards greater accountability of multinationals for environmental failures (see 'Nigerian Villagers Win Right to Pursue Shell in English Court', Sandra Laville & Emmanuel Akinwoto, *The Guardian*, 13 February, 34).

30 See: Jasper Jolly, 'Climate Crisis: FTSE Giants Fail to Disclose Their Carbon Footprint', <https://www.theguardian.com/business/2020/dec/13/climate-crisis-ftse-giants-fail-to-disclose-their-carbon-footprint> (accessed 13 December 2020).

31 And with the USA the largest oil-producing country in the world. See: <https://www.statista.com/topics/1783/global-oil-industry-and-market/> (accessed 7 July 2020).

32 The figures for 2018 were $3.5 trillion, so maybe already 2019 was seeing a downward trend and production for 2020 is predicted at $1.47 trillion – see: <https://www.iea.org/reports/oil-2020> and <https://www.cnbc.com/2020/04/30/coronavirus-creating-1-trillion-revenue-loss-for-oil-and-gas-companies.html> and see also: <https://en.wikipedia.org/wiki/List_of_largest_oil_and_gas_companies_by_revenue >. Saudi Arabia, the USA (both 13%) and Russia (12%) are the three largest oil-producing countries, see: <https://www.statista.com/topics/1783/global-oil-industry-and-market/> (all accessed 07 August 2020).

33 See: <https://www.cnbc.com/2020/04/30/coronavirus-creating-1-trillion-revenue-loss-for-oil-and-gas-companies.html accessed 07/08/2020>. And more recently see how a major investment firm, the Norwegian hedge fund Storebrand, has divested itself of Exxon, Chevron and Rio Tinto stock. See: <https://www.theguardian.com/environment/2020/aug/24/major-investment-firm-dumps-exxon-chevron-and-rio-tinto-stock> (accessed 24 August 2020). Also, a recent report indicated that, even though only 10% of electricity utility companies are prioritising renewable energy, nonetheless the majority of them are clustered in Europe (little shifts, but…). See: <https://www.theguardian.com/business/2020/aug/31/only-one-in-10-utility-firms-prioritise-renewable-electricity-global-study> (accessed 31 August 2020). And why, given that investors generally require solar and wind energy projects to make a

return of 10% because they are low risk, and require returns of 40% from fossil fuel investments because they are high risk, should they continue predominantly to take the latter route, when it seems self-evident that renewable energy systems are the ones to invest in? And the answer is of course because – all risks aside – there are at least some decades to go before coal (in particular) runs out, so there is still money to be made.

34 For more on this gloomy news see: <https://www.theguardian.com/environment/2020/dec/02/world-is-doubling-down-on-fossil-fuels-despite-climate-crisis-un-report>; <https://www.theguardian.com/business/nils-pratley-on-finance/2021/feb/02/lockdown-hit-bp-is-still-at-a-very-early-stage-of-its-required-reinvention> (both accessed 3 February 2020).

35 See his book *Confessions of an Economic Hit Man*, Oakland CA, Berrett-Koehler Publishers (2004).

36 See: <https://www.landsvirkjun.com/company/powerstations> (accessed 11 August 2020).

37 For further details see: <https://grapevine.is/news/2013/03/13/deterioration-of-lagarfljot-was-foreseen/> (accessed 18 August 2020).

38 See: <https://grapevine.is/news/2013/03/14/impact-of-karahnjukar-was-known-former-progressive-minister-of-environment-stands-by-the-project/> (accessed 11 August 2020).

39 For recent ALCOA contracts with the US military see: <https://www.businesswire.com/news/home/20160303005234/en/Alcoa-Awarded-50-Million-Contract-U.S.-Army> (accessed 12 August 2020).

40 One suspects the choice of *Draumalandiö* as the title of the documentary is totally intentional since *draumr* refers to one who is in an awakened state of mind but distracted from reality and oblivious to all practical concerns.

41 See: <https://balkanriverdefence.org/the-undamaged/> (accessed 12 August 2021).

42 For details on this Danish-based project which seeks to popularise a global debate on poverty and inequality see: <https://www.thewhy.dk/projects/why-poverty> (accessed 12 August 2020).

43 For more details see this interesting study: <https://resourcegovernance.org/sites/default/files/nrgi_NMC_English.pdf> (accessed 13 August 2020).

44 In: <https://www.idfa.nl/en/film/fe086808-4690-4544-9b11-223f1132740b/stealing-africa> (accessed 12 June 2020). In 2011 Glencore mounted the biggest flotation on the British stock exchange. According to *The Guardian* 'Glencore's market share is so large that it recorded revenues of $145bn (£89bn) last year and the flotation value is £37bn. Glasenberg, owning around 16%, will instantly become one of the world's richest men'. And because 'the company will leap straight into the FTSE 100 index, those investments that track the UK's biggest quoted companies will be obliged to buy its stock, with the result that the company's fortunes will have a bearing on the pension funds of millions of people'. <https://www.theguardian.com/business/2011/may/19/rise-of-glencore-commodities-company> (accessed 14 August 2020).

45 An equally strong exposé of this dreadful exploitation comes in a French television documentary by Audrey Gallet and Alice Odiot, *Zambia: Good Copper Bad Copper* (2011) broadcast on France 5.

46 This isn't the first time I address this point (see footnote 9), but every year multinationals' wheeling and dealing deprive developing countries of around $160 billion in taxes, which explains how 'Third World' countries are victims of the processes of globalisation.

47 In the documentary, there is the very forceful presence of Eva Joly, Chair of the EU Committee for Development, who repeatedly attempts to enforce transparency of practice. Ultimately to little effect.

48 There is every evidence that the EIB continues to support infrastructure projects linked to human rights violations in Africa and Asia. See: <https://www.theguard

ian.com/global-development/2020/nov/09/eu-bank-supports-projects-linked-to-human-rights-violations-ngos-claim> (accessed 9 November 2020).

49 Whilst this is alluded to in *Stealing Africa*, I'd like to draw your attention to the EIB's response to the complaint, see: <https://www.eib.org/en/press/news/update-on-the-status-of-the-eib-loan-for-the-mopani-copper-project-zambia> (accessed 17 August 2020).

50 *Stealing Africa* informs us that, amongst others, the Church of England and Norway are shareholders in Mopani Mines.

51 See: <https://www.business-humanrights.org/en/zambia-mning-impacts-on-communities-near-town-of-mufulira> (accessed 17 August 2020).

52 In fact a couple of publications expose Banda's foreign-affairs dealings as clearly corrupt and deliberately neglectful of the Zambians' poverty and indigence: <https://www.thelocal.ch/20110920/1224> and <https://www.theafricareport.com/8480/zambia-mining-seeking-an-international-stake/> (both accessed 17 August 2020). However, there was a glimmer of justice in May 2020 when the Zambian Supreme Court showed its teeth and fined Glencore $13 million for tax avoidance. See: <https://www.business-humanrights.org/en/latest-news/zambia-court-fines-glencores-mopani-copper-13-million-for-tax-avoidance/ (accessed 21 May 2020). Interestingly, though surely not connected, Glasenberg resigned as CEO, leaving Glencore to the next generation.

53 For an in-depth study of the levels of mass-scale looting and other quite shocking corrupt practices in relation to the DRC's wealth in minerals see: <https://en.wikipedia.org/wiki/Mining_industry_of_the_Democratic_Republic_of_the_Congo> (accessed 19 August 2020).

54 <https://www.theatlantic.com/international/archive/2014/06/gorillas-guns-and-oil-a-journey-inside-virunga-national-park/373305/> Nicholas Slayton interviewer (accessed 12 August 2020).

55 <https://www.latimes.com/entertainment/envelope/la-et-mn-virunga-20150205-story.html> Janet Kinosian interviewer (accessed 12 August 2020).

56 <https://www.latimes.com/entertainment/envelope/la-et-mn-virunga-20150205-story.html> Janet Kinosian interviewer (accessed 12 August 2020).

57 In further researching this issue of government officials selling off areas for oil prospecting in the eastern region of the DRC, I came across a number of informative articles which further enlighten this tale. Most of the oil-rich areas are in eastern DRC (which is also home to Virunga National Park), but the conditions for drilling (in thick equatorial forests) and exporting are both difficult and dangerous, so any signed agreements are unlikely to be fulfilled. There are no roads to facilitate transporting the oil to the west coast thousands of miles away. As for going to the east, there is currently no pipeline, although there is a planned construction (in which the French oil company Total has a shared interest) of a Uganda–Tanzania crude oil heated pipeline (897 miles long through major chimpanzee and elephant reserves, crucial biodiversity hotspots such as Lake Albert, see: <https://e360.yale.edu/features/a-major-oil-pipeline-project-strikes-deep-at-the-heart-of-africa>). However, the DRC does not have good relations with Uganda so if the pipeline is built it is not at all certain the DRC-based oil prospectors would be allowed to access it. The government does not, therefore, have any interest in developing this sector. The thing of value is not the oil itself but the signed contracts which line the politicians' pockets. The oil sector is not there to be developed, but to serve as a reward to regime loyalists. The contracts are sold with the knowledge there is little risk of oil exploitation. Indeed, politicians sign them and then tear them up, or sell the same blocks to different buyers (who in turn might sell them on for profit to other buyers). See: <https://www.privacyshield.gov/article?id=Congo-Democratic-Republic-Oil-and-Gas>; <https://qz.com/africa/1567958/dr-congos-kabila-corruption-hampers-oil-industry/>; and for an original twist in the tale, <https://www.reuters.com/article/congodemocratic-oil-gertler/update-3-israeli-billionaire-finds-3

-bln-barrels-of-oil-in-congo-idUSL6N0QD40I20140807> (all accessed 19 August 2020).

58 <https://www.theatlantic.com/international/archive/2014/06/gorillas-guns-and-oil-a-journey-inside-virunga-national-park/373305/> Nicholas Slayton interviewer (accessed 12 August 2020).

59 <https://www.indiewire.com/2014/11/springboard-virunga-director-orlando-von-einsiedel-on-why-avatar-was-a-huge-inspiration-68278/> (accessed 19 August 2020).

60 Ibid.

61 See: <https://deadline.com/2015/02/virunga-orlando-von-einsiedel-leonardo-dicaprio-oscar-nominated-documentary-1201371343/>; and <https://www.standard.co.uk/go/london/film/the-gorillas-leo-and-me-orlando-von-einsiedel-interview-10029096.html> (both accessed 20 August 2020).

62 <https://deadline.com/2015/02/virunga-orlando-von-einsiedel-leonardo-dicaprio-oscar-nominated-documentary-1201371343/> (accessed 20 August 2020).

63 As of 2021, Total is still determined to fulfil this project of the Uganda–Tanzania pipeline; as oil becomes scarcer it is quite possible that they will get their way and thrust this ecological disaster across numerous national parks to the detriment of natural ecosystems.

64 This visionary man described himself as a 'comprehensive anticipatory design scientist'. He dedicated his life to making the world work for all of humanity, beginning with the principle of making 'more with less'. See: <https://www.bfi.org/about-fuller > (accessed 23 August 2020).

65 See: <https://www.climatepartner.com/en/carbon-offset-projects/hydropower-virunga-dr-congo>. See also that a second hydropower plant in Virunga is on the cards <https://wwf.panda.org/?213450/Construction-begins-on-new-Virunga-hydropower-plant> (both accessed 3 April 2021).

66 Here is just one example of network misinformation taken from the UK government's think tank The Global Warming Policy Foundation's website which shows how their language indirectly denies that global warming is overwhelmingly scientifically proved (which of course it is). This is the sort of rhetoric they use around the issue of climate change: 'while it is open-minded on the *contested science of global warming*, it is deeply concerned about the costs and other implications of the policies currently being advocated'. See the article in full <https://www.theguardian.com/environment/2020/sep/02/ground-zero-of-lies-on-climate-artists-protest-at-london-thinktanks> (accessed 2 September 2020).

67 Raworth quotes Janine Benyus, a leading proponent of biomimicry, who proposes that it is something we should actively pursue: 'let's learn to halt our industrial "exhale" of carbon pollution, and by mimicking plants, learn to "inhale" carbon dioxide into our products and store it for centuries in rich agricultural soils' (Raworth, 2017: 218).

68 See: <https://www.ellenmacarthurfoundation.org/assets/downloads/publications/Ellen-MacArthur-Foundation-Towards-the-Circular-Economy-vol.1.pdf> (accessed 26 June 2020).

69 I can at least draw your attention to two studies you might enjoy (Bonner, 2007, and Hughes, 2014).

70 See: <https://pdfslide.us/documents/the-4th-revolution-energy-autonomy.html> (accessed 26 June 2020). Even the film itself is democratically funded by crowd funding.

71 In his book *Energieautonomie* (2005) Scheer challenges the 'unbroken power of our handed-down energy thinking' and demonstrates ways in which the complete transition to renewable energy can succeed. See: <https://pdfslide.us/documents/the-4th-revolution-energy-autonomy.html> (accessed 26 June 2020).

72 On the issue of batteries, we hear from Professor Maria Skyllass-Kazacos, the inventor of vanadium flow battery systems for electricity storage. Whilst safe and dependable and efficient, nonetheless, they rely on lithium as do all storage

batteries, so they are not a complete solution. There are moves towards a new kind of storage, this time in bricks – using fired red house bricks as supercapacitators. See: <https://www.theguardian.com/environment/2020/aug/11/powerhouses-nanotechnology-turns-bricks-into-supercapacitor-batteries> (accessed 11 August 2020).

73 <https://pdfslide.us/documents/the-4th-revolution-energy-autonomy.html> (accessed 26 June 2020).

74 This was in 2010. By 2019 this had risen to 97% of borrowers being women. And banks receiving 94% of their micro-loans back. See: <http://www.grameen.com/breaking-the-cycle-of-proverty/> (accessed 12 September 2020).

75 Willenbacher and Jung are visionaries with a holistic approach committed to 100% renewable energies. See: <https://www.renewablepress.com/energy/press-release-3035-juwi-to-expand-company-headquarters-en> (accessed 12 September 2020). The juwi Group has subsidiaries in 12 countries and has established over 140 solar, wind and bioenergy plants over the world. See: <https://www.juwi.com/company/references/reference-detail/artikelansicht/burgos-solar-i> (accessed 12 September 2020).

76 Increasingly, Elon Musk stands out as something of a charlatan when it comes to ecological practice. As a billionaire, he seems more interested in accruing wealth – founding and selling on companies: beginning with software Zip2, sold to Compaq; then, online banking eventually merged into PayPal and sold to eBay. Now, as head of SpaceX, his dream is to reduce the expense of space travel! None of these endeavours has much to do with regenerative practices.

77 See: <https://www.virta.global/global-electric-vehicle-market?__hstc=51530422.3eda98961dfce0b976b1f9be92ef304e.1620462271777.1620462271777.1620462271777.1&__hssc=51530422.12.1620462271777&__hsfp=3552636164&hsutk=3eda98961dfce0b976b1f9be92ef304e&contentType=standard-page#one> (accessed 8 May 2021).

78 See: <https://www.vox.com/future-perfect/2019/1/15/18182167/microcredit-microfinance-poverty-grameen-bank-yunus> (accessed 7 May 2021).

79 There is plenty of evidence that oil firms knew decades ago fossil fuels posed grave health risks. See:<https://www.theguardian.com/environment/2021/mar/18/oil-industry-fossil-fuels-air-pollution-documents> (accessed 19 March 2021).

80 Birol evidences his complete indifference to the impact of CO_2 emissions. Consider for a moment that China is responsible for 30% of greenhouse gas emissions, whereas the African continent is responsible for a mere 3%. Twenty countries in the world, mostly wealthy nations, are responsible for 80% of the world's CO_2, the consequences of which most severely affect poorer nations – especially in the African continent.

81 Fatih Birol quoted in: <https://www.theguardian.com/environment/2021/apr/20/carbon-emissions-to-soar-in-2021-by-second-highest-rate-in-history> (accessed 20 April 2021).

82 Fatih Birol quoted in: <https://www.theguardian.com/environment/2021/may/11/global-renewable-energy-industry-grew-at-fastest-rate-since-1999-last-year> (accessed 11 May 2021).

83 In follow-up articles we learn that Badr is caught stealing equipment and Rafea's husband is accused of embezzling money from the project (having already been jailed for drug-smuggling). See: <https://www.pbs.org/independentlens/blog/amid-setbacks-the-solar-mamas-are-undeterred/> (accessed 20 September 2020). This was the case in early 2013. Since then a further interview with Rafea shows just how much the empowerment she felt from the course has enabled her to transcend these earlier setbacks. See: <http://www.trustconference.com/news/i/?id=a7149f4e-3eb7-4b77-a9eb-74c4d7f4a382> (accessed 22 September 2020).

84 For example, Raouf Dabbas, when on the phone to Rafea's husband in an attempt to bring him around, apologises to Rafea for having to use the tactic of 'speaking man to man' to Alian so that he can agree without losing face to her returning to India.

85 See Fatih Birol's recent comments in *The Guardian* ; for example https://www.theguardian.com/environment/2020/jun/18/world-has-six-months-to-avert-climate-crisis-says-energy-expert; https://www.theguardian.com/environment/2021/feb/01/world-kick-coal-habit-start-green-recovery-iea-fatih-birol; and https://www.theguardian.com/environment/2021/may/18/no-new-investment-in-fossil-fuels-demands-top-energy-economist. He has also been speaking out, frequently, about the dangers to the environment in *The Financial Times*.

2

WASTE … MEANS … MONEY … AND SUFFOCATING THE EARTH – THE HUMAN PRACTICE OF POLLUTION

The world's present rate of garbage output is not only unsustainable but, in many cases, is life-threatening to flora and fauna, humans and other-than-humans alike. Garbage itself is unsustainable, not just because of the amounts produced, but also because of our current inadequate disposal facilities, the inept and corrupt practices by those who stand to make profit from waste disposal and the failure of the judiciary to successfully prosecute criminal waste-disposal practices (exercised by waste-management business firms, as much as by any criminal bodies). But there is equally the fact that, beyond these malpractices and the impact of industrial waste on the environment, we each contribute to this problem. Figure 2.1 is a mapping of our wasteful ways (in kilograms per person per year).[1]

Raworth points out, in *Doughnut Economics*, that the current pattern of waste production occurs because the mainstream economic growth model is 'an *open system* with constant inflows and outflows of matter and energy' (2017: 74). But it is a flawed design, for the following reasons. The 'economy depends on the Earth as a *source*', however, beyond 'harvesting renewable resources like timber, crops, fish and freshwater', the economy also extracts 'finite resources like oil and minerals' (212). Moreover, the economy uses 'the Earth as a *sink* for its wastes' which include domestic as well as industrial waste – all of which is manufactured by humans (212). This economic behaviour ignores the fact that the planet is a *closed system* 'which thrives by continually recycling life's building blocks such as carbon, oxygen, water, nitrogen and phosphorus' (212). In our Anthropocene era, our *oikos nomos* – that is, our economics or planetary management – consists of taking, making and dumping. And what makes this misguided practice almost impenetrable is that whilst, indeed, economics is the big driver behind this state of affairs, colluding with it are the legal and political institutions – what I referred to in my earlier book *Film Ecology* as a magical triangulation or striangulation and which works as follows:

DOI: 10.4324/9781003130246-2

> economic, legal and political institutions work hand in glove to organise society for their own benefit. It's a magical triangulation, a striangulation you might say. It is, after all, as Acemoglu and Robinson state, 'the political institutions that determine what economic institutions a country has.' (2013: 43) It is also the case that political institutions pass the laws that govern a country and so determine the legal institutions that a country has. As a result, the way that political power is exercised is as much influenced by economic as it is by legal institutions. This dynamic makes clear how the regulatory system of a country is monopolised by a narrow elite, and how the inevitable outcome is, at best, a diminished state and, at worst, a lack of equality and political rights for the majority. Here's how the loop whizzing round the triangulation works: 'political institutions enable the elites controlling political power to choose economic (and legal) institutions with few constraints or opposing forces. They also enable the elites to structure future political institutions and their evolution.' (81) Trying to fight against this evidence is a bit like trying to cut off Hydra's head. Everytime a small victory for justice is won and Hydra's head is lopped off, just as quickly Hydra produces another — if not several heads — and the battle begins all over again.
>
> *(Hayward, 2020: 74, parenthesis mine)*

In the nine eco-docs pertinent to the issue of waste and pollution, some of the major consequences of our excessive consumption habits and throwaway culture are brought into full focus (see Table 2.1). Over-consumerism and its many impact factors, urbanisation and its concomitant pollution, plastic and other toxic-waste disposal, the dangers posed by toxins in everyday life, largely produced by chemical companies – all these issues are explored in detail in these eco-docs whose purview includes the Americas, the Middle East and Europe. Air, water and land pollution means that there is virtually no safe space left for the human species – amongst others – to breathe. And as we suffocate ourselves to death, so too we deprive the Earth of its own oxygen needs. As Candida Brady, director of the eco-doc *Trashed*, correctly asserts, 'the whole of humanity's way of life has got out of control'.[2]

TABLE 2.1 Table of eco-docs about pollution from domestic, medical and industrial refuse

I. Water	*II. Air*	*III. Land*
1. *Before Vanishing* Fr/Syria (2005)	2. *La bambina deve prendere aria/Baby Needs Some Fresh Air* It (2008) 3. *Clouds of Smoke* Albania/UK (2007) 4. *No Pressure* UK[3] (2010)	5. *Biùtiful cauntri/Beautiful Country* It (2007) 6. *Una montagna di balle/Mountain of Lies* It (2009) 7. *Polluting Paradise* Ger (2012) 8. *Trashed* UK (2012) 9. *Waste Land* UK/Brazil (2010)

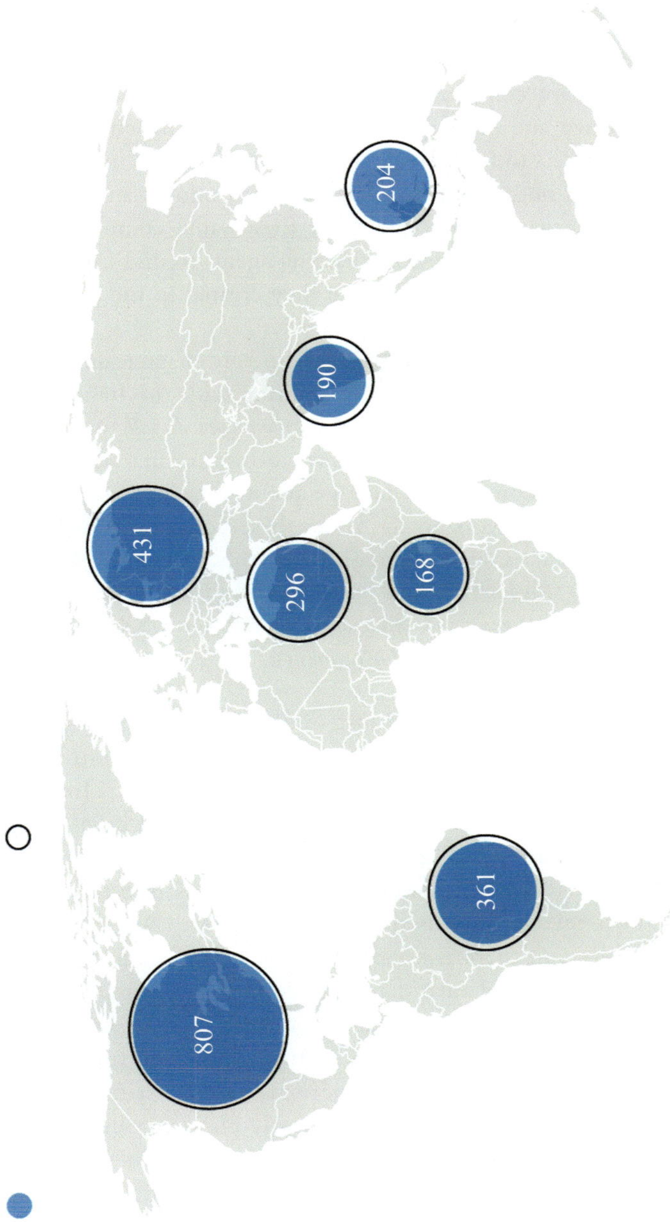

FIGURE 2.1 Mapping our wasteful ways: kilograms of solid waste generated per person per year.

In this Anthropocene era, none of us remain unaffected by the domestic and industrial practices of waste production and disposal: children die from exposure to the urban environment, industrial waste induces health problems on a worldwide scale, the cost of both permitted and illegal waste management to environmental and human health is equally immense. Waste management, however, is big business which banks are only too happy to help finance. For, not only do refuse-management companies profit from processing the waste produced by humans, industry and corporate businesses, they also benefit hugely from governmental contracts to both create landfills and build incinerators. And the money-making does not stop there. When landfill odours and waste-leachate occur, due to poor management practices, these same companies frequently end up delivering the (often untested) cleaning agents to disguise the smell and decontaminate the environmental damage they created in the first place (once again, often subsidised by governmental agencies).

This money game will be a key emphasis in this chapter, and the primary focus will be on four of the eco-docs, beginning with the Italian trilogy (*La bambina deve prendere aria, Biùtiful cauntri, Una montagna di balle*) where I explore the disastrous consequences of deregulated waste-disposal practices. The chapter then concludes with a more upbeat message about waste processing from *Waste Land*. To provide a framework to the issues these pollution eco-docs address, however, here, for a starter, are some flash statements.

First, some sobering facts from Trashed *brought to us by Jeremy Irons during his trash-odyssey across the world's oceans and numerous countries (the USA, Asia, the Middle East, and Europe). Here is a summary of his findings:*

> In **Lebanon**, along the coast just outside Beirut, the Mediterranean Sea laps up against the stinking household trash-mountains, 40 metres high, rendering beaches unsafe and polluting the water. In **European**, **Chinese**, **North** and **South American** towns and cities, most communities live within 2–3km of a landfill site or incinerator. Both present problems to health. Land-fill leeches toxins into the air and soil; incinerators release dioxins into the atmosphere. We all have dioxins in our bodies and there is no safe dosage (as the example of the use of Agent Orange in **Vietnam** demonstrates). When dioxins get to colder parts of the world they condense. The **Arctic** is the most contaminated place on Earth and, as it melts, so toxins get released into the environment. All **oceans** are polluted. There are five centrifugal areas in the oceans (5 Gyres) where our waste, mostly plastic, collects — thus, all phenomena of sea-life absorbs the micro-plastic fragments. This plastic is not inert, but absorbs other compounds such as dioxins in the water. So the substance's concentration magnifies progressively as it moves up a food chain from small to larger sea-life and eventually onto our plates (a process known as biomagnification). The documentary ends with indications how to resolve our deregulated waste problem.

Second – in relation to Clouds of Smoke*, the director Fatmir Tirziu informs us:*

> **Albania** is one of the most polluted countries in Europe. According to the World Health Organisation (WHO) this country is facing a humanitarian catastrophe due to growing pollution caused by poisonous gases. According to the responsible person for the laboratory of industrial toxicology in Elbasan, Beqir Kila, the scientists 'have discovered nickel in quantities several times higher than the acceptable norm and the ground is polluted to the depths of about 35 centimetres' (Spaho, 2007: http://enrin.grida.no/). These gases are feeding the atmosphere, are leading to climate change effects and polluting not only the country, but also all the area around it. Albania's 3.5 million people still don't have a single wastewater treatment plant, toxic waste disposal facilities or sanitary landfill. The pollution was created mostly by the fumes that the old factories released and of burning of the waste in the waste fields. Another problem is the burning of the city's waste. It is easy to find the way to the city's waste dump as you can easily follow the trail of smoke to the place. This is because the rubbish is almost always being burned.[4]

Third, from the press release for Polluting Paradise*:*

> The mild moist breeze sweeping in from the Black Sea feeds the evergreen, precipitous land that its inhabitants have transformed into a garden: A carpet of lush green, round tea bushes stretching out between pines and rhododendrons; the livelihood of the 1700 inhabitants of Camburnu. For generations the village in the northeast of **Turkey** has lived on tea cultivation on the slopes and fishing in the sea. But this peace has been threatened since the government decided to build the largest garbage dump of the eastern Black Sea region in an area located above the village. 'Our village is being sacrificed,' says the mayor. He fears that soon no one will want to buy any more tea or fish from Camburnu, because the garbage dump above the village will poison the ground water and pipes will flush the waste into the sea.[5]

Finally, with regard to the context of the Italian trilogy, Biùtiful cauntri*,* La bambina deve prendere aria *and* Una montagna di balle*:*

> In the normal parlance, the term 'Land of Fires' indicates an area in Campania, in the south of **Italy** where, systematically, since the end of the 1980s, toxic wastes have been illegally burnt and buried. Organised crime plays a significant role in illegal waste trafficking, however, organised mafia-like criminals are not the only players. Indeed, a more substantial explanation of the phenomenon is articulated around *the interplay of organised mafia-like crime, businessmen, firms and administrative officers.* The

> weak enforcement-power at both national and regional levels has also been used to explain this widespread illegal situation. Environmental laws and, in particular, the sanctions related to misconduct in the waste management are not stringent enough and do not at all represent a deterrent for criminals. Most of the violations are considered misdemeanours. According to the Italian Anti-mafia National Prosecutor, firms commit most of the waste-related environmental crimes, often without any direct connection with organised group mafia-like; however there is no norm to convict juridical person. The consequences of such weak enforcement have been the increasing number of contaminated hot spots in the Campania region and profound concerns for the health consequences of such contamination.[6]
>
> *(emphasis mine – I would also draw your attention to Anita Angelone's excellent article on this trilogy[7])*

Let us now take a closer look at this trilogy of Italian eco-docs which provides such an exemplary instance of how the self-serving money game of «garbage-management» works.

2.1 The money game – triangles within triangles: *Biùtiful cauntri, La bambina deve prendere aria* and *Una montagna di balle*

These three documentaries emerged within one year of each other (2007–9), and all address the scandalous waste-problem situation in the Southern Italian region of Campania that has progressively worsened since the authorities first declared it a 'state of emergency' in 1994. Despite repeated promises to do something, despite appointing numerous special government commissioners in charge of waste management, despite numerous demonstrations, despite meetings held by local councils and municipalities decrying the state of affairs, by the time of these eco-docs' release – 13 to 15 years after the 1994 'state of emergency' – little to nothing had actually taken place to remedy this problem. Nor, indeed, some 20 years later, for, in 2015, an EU-funded report found that:

> In the last 20 years, tens of thousands of citizens have been denouncing the disastrous results of the authoritarian implementation of the urban waste plan, the complete lack of intervention of the authorities against the illegal waste dumping, the lack of effective remediation of the thousands of contaminated sites and, last but not least, the superficial and unsatisfactory responses coming from the national healthcare system to the health alarm of the Campania population.[8]

And yet millions of euros have poured into resolving this crisis, with very little to show for it. Just to cite two examples. An initial €174 million in 1997 was paid out by the government to decontaminate existing polluted areas; to date no single site has been decontaminated (see end of *Biùtiful cauntri*). From 1998 onwards, 90% of public funding intended for renewable energies (€6 million yearly) went into resourcing incinerators to be built by the privately owned waste-disposal company FIBE of which only one (of a projected seven) was completed in 2009 (see *Una montagna di balle*). Which begs the question, where has this money gone? Taking the documentary evidence of our three eco-docs together, this is how the money game works:

1. ***First***: Campania is composed of five provinces (Avellino, Benevento, Caserta, Naples and Salerno) each with its own council and president albeit with limited powers and answerable to Naples which is the seat of the Regional Council of Campania. Local waste is the responsibility of each province. Beginning in the 1980s, the 80 Campanian landfills, originally intended just for local waste, began receiving refuse (including industrial and medical waste) from Northern Italy. In the money game, this constitutes a first financial transaction – *a trade in waste*, the North pays the South. This not only puts control, over the type of waste deposited, beyond the purview of the local authorities, it also means, somewhere, people in local authority are making money on the side.

   Outcome: by the mid-1990s there is no space left in the 80 landfills for the local waste, and so waste accumulates in the streets of Naples and its surrounding suburbs as well as other towns, especially Caserta, and *comunes* in the Campania region. The increasing mounds of rubbish result in poisoning the air.
2. ***Second***: all original landfills reach capacity, so more fill spaces are required.[9] In 1994, a state of emergency is declared by the national government (under Berlusconi) and implemented by the Regional Council (i.e., Naples) in the form of a special Commission. So now the waste problem is under the purview of the Regional Council of Campania (and out of the hands of the local provinces). It takes the newly established Commission four years to pursue any action: waste-management businesses are invited to tender waste-disposal proposals for refuse-driven-fuel (RDF) plants which include the use of incinerators. In 1998, FIBE, a consortium of Salini-Impreglio, wins the bid, based on its low cost rather than its proposed use of technology to manage the waste crisis. The promised seven incinerators are extremely slow in coming into practice.[10] To compensate for this slow progress, however, FIBE is given the power by the Commission's authorities to acquire sites for its facilities without local consultation. So instead of pursuing the construction of its incinerators (for which it is receiving public funding, remember) FIBE continues its practice of storing waste in plastic-bound bales (known as *ecoballe*), which are then piled into huge mounds (often adjacent to farmland) and covered with yet more plastic sheeting with the promise that

they will be processed by incinerators (once up and running). A second financial transaction occurs – *a legal privatisation of waste for profit.*

Outcome: merely storing unrecycled waste – brought from the North as well as the local region of Campania – in bales is a far more profitable process for FIBE than the building of incinerators and processing the fall-out. FIBE pockets money, and its promises to invest its profits and public funding into incinerator plants come to nought. Meantime, the leachate from these bales is poisoning the soil and local water sources.

3. ***Third***: during this slothful activity on the part of the Commission, Campania's Regional Council and FIBE, illegal dumping becomes common practice. The running of this side of the waste business, in particular toxic and radioactive waste, is run primarily by the Camorra (the local mafia).[11] The authorities meanwhile do nothing to curb this. Since the late 1990s, some 1200 illegal dumping sites in Campania alone are the outcome of this laissez-faire – particularly in the eastern agricultural region, but also to the west of Naples where untreated waste is burnt, in an area that has become known as the *terra del fuochi*/land of the pyres. The Camorra, labelled the *ecomafia*, profits from illegal waste-dumping which is second only to that earned from their drugs and arms trafficking and is infinitely less risky, since it is more or less condoned by the authorities, helped by backhanders of course. Thus, a third financial transaction occurs – *an illegal privatisation of waste for profit.*

   Outcome: the triangle of the money game is complete. Both legal and illegal waste-dumping is subject to no real administrative control. All three parties enrich themselves (business, council, the Camorra). Meantime, to the west and north of Naples air pollution from the land of pyres and uncollected garbage in the streets increases; to the east, in the agricultural area, leachate intensifies, toxic waste is worked into the land under the guise of fertiliser (amongst other practices which include burning industrial waste mixed with textiles). This area, known formerly as *campania felice*/beautiful or happy country, has been turned into trash country, now known as the 'triangle of death'. The farmers' land, crops and livestock are decimated, and the local residents' health deteriorates.

This then is the profit-making waste-dumping game to which the three Italian eco-docs speak. *Biùtiful cauntri* (Esmeralda Calabria and Andrea D'Ambrosio, 2007, Italian Golden Globe winner) provides us with the biggest portrait of the problem confronting the inhabitants of the Campania region. It places its investigation, led by the then-Environmental Counsellor Raffaele Del Giudice, in three key areas: to the west of Naples and the land of pyres with the *comunes* of Giulignio and Villarica and that area's RDF plant, Taverna Del Re (an FIBE plant with no incinerator, of course); to the east and the so-named triangle of death (the *comunes* of Acerra, Nola and Marigliano) with particular focus on the legal and illegal waste-dumping in Acerra; and, finally, to the south and more illegal dumping by the Camorra of toxic waste in the national park of Vesuvius. *La bambina deve prendere aria* (Barbara Rossi-Prudente, 2008) is set in Caserta, 26 km north of Naples, where – as with the sister city – mountains of rubbish have accumulated. We follow the meanderings of a mother pushing her baby in a stroller through stench-filled streets as she explains to her child that her doctor has told her she *must/deve* take her baby out to get fresh air! Would that she could... The mother is the documentary filmmaker, Rossi-Prudente, and this odyssey through the streets with her daughter, Lucrezia, brings her into contact with numerous witnesses (some, ordinary citizens young and old; others, experts in their fields – doctors and architects amongst them) whose responses to the garbage crisis she records. Finally, *Una montagna di balle* (Nicola Angrisano, 2009) focuses on the mountain of rubbish (*ecoballe*) accumulated in the Acerra area by FIBE at its various plants and the 'mountain of lies' (the other meaning of the eco-doc's title) surrounding FIBE, the municipality and the government's promises to deliver a solution to the dreadful contamination of air, soil and water that has ruined the local farmers' and other food producers' businesses in an area formerly known for its agricultural plenitude.

Campania owes the richness of its soil to its geographical location at the crossroads of the two major tectonic fault lines (west to east and north to south), making this region the only self-sustaining agricultural area in Southern Italy.

> Except for the volcanic region around Naples, farming in southern Italy is exceedingly difficult because limestone forms the basement rock and the soil is generally quite poor. But the region around Naples, which includes Mount Vesuvius, is very rich mainly because of two large eruptions 35,000 and 12,000 years ago that left the region blanketed with very thick deposits of tephra which has since weathered to rich soils.[12]

Both *Biùtiful cauntri* and *Una montagna di balle* provide visual evidence of the mere 15 (1994–2009) years it has taken for a significant part of that region's nurturing munificence to be eradicated. Shepherds relate their story and show us proof of their losses: sheep and their lambs dying of dioxin poisoning because of toxic-waste leachate into their pastureland from the adjacent landfills and accumulated *ecoballes*. The same is true for the farmers and their buffalo livestock. The food produced (dairy, meat, vegetables and fruit) is so contaminated it, too, is a source of harm and so cannot be sold, nor indeed eaten by its producers. Estimates stand at between 14% and 32% of Campania's agricultural land being so polluted that, as one farmer in *Biùtiful cauntri* puts it, 'this area is being wiped off the map' and 'we are dying'.[13]

As for city and town life, *La bambina deve prendere aria* offers a similarly distressing story. Rossi-Prudente explains to her daughter that 'garbage permeates our everyday behaviour; the garbage enters your head', it seeps into your clothes.

Not just clothes, the mother continues, but buildings also; thus, 'mayors sell off land-filled land before the 30-year ban is completed, so buildings are put up on toxic waste sites', with buildings themselves at times being made of the illegal waste the Camorra has buried in concrete. Such is the violence of this garbage's presence, remarks the architect Beniamin Servino, there is no recognition of a public space anymore where people can gather, which is their civil right. Instead, people lead parallel lives, many seemingly inured to this malodorous presence – that is until they rise up and protest, when they become a *comune* once more, albeit short-lived, either because there is no response from the council, or demonstrators are dispersed by police force. We learn, from tests run by paediatrician Gaetano Rivezzi, how this violent presence also enters the maternal body, in the form of alarming levels of dioxins in their breast milk, this time absorbed from the burning of unrecycled waste. Already the pollution has assailed the heath of children and cancers are increasing amongst young people. Now, equally disturbing, babes-in-arms are ingesting unsafe produce from what should be the very site of maternal care.

Instead of the nurturing, which is implicit in the maternal body, or the protection and care intrinsic to the good shepherd, what is present in all three eco-docs are examples of the total abuse of the 'Right to Health' guaranteed by the Italian Constitution. The assault on the land, on the inhabitants and on other-than-humans is matched only by the disregard for the civil liberties of those affected and the contempt shown, towards the locals' sense of outrage, by both the media and the perpetrators of the environmental damage. The European Commission has repeatedly reprimanded Italy for flouting the Hazardous Waste Directive and the Landfill Directive.[14] But, instead of responding with solutions, the Italian government, especially under Prime Minister Berlusconi (1994–5, 2001–6 and 2008–11), has taken an increasingly authoritarian turn in waste policies. Thus, Berlusconi has issued a series of states of emergency, reopened a number of quarries in Campania that had been closed because they were deemed unsafe and, in response to citizens' grievances and demonstrations against waste-facility siting in Campania, instituted, in May 2008, the 'Law of Felony against Protesting'.

In both *Biùtiful cauntri* and *Una montagna di balle* we witness examples of the progressive militarisation of landfills and waste-plants and increased police repression of protests. In *Biùtiful cauntri*, Raffaele Del Giudice visits a number of sites and is prevented from filming and entering (even though he is Naples' Environmental Councillor) by guards employed to keep citizens away. In *Una montagna di balle* we see protestors being met with police violence. The following statement from one farmer interviewed in *Biùtiful cauntri* sums this up as follows:

> Nothing can be done about it because those people just don't care about the damage they do to others. The Prefects, the authorities, they are the ones with power, they have the power to repress crime. They came to see us when the dump was being stormed by the local population. But they

> came with a battalion of soldiers and police—a battalion, a 1000 men—so it was impossible to understand what was going on … as things stand now, life will be impossible in 15–20 years from now.

Time and again we see examples of this indifference and contempt within this triangle of power: the authorities, the businessmen and the Camorra. *Biùtiful cauntri* serves up instances of the authorities' disregard, the most explicit of which is when the Commissioner for the State of Waste Emergency, Guido Bertolaso, turns up at Villaricca's waste dump (Cava Riconta) at the behest of the local protest group headed by Del Giudice. When met with their questions and demands, his response is silence and he walks away. Indeed, no one in power is listening. With regard to big business, both *Biùtiful cauntri* and *Una montagna di balle* expose the consequences of FIBE having an exclusive monopoly on waste management for 15 years, during which time it 'took the farmers' land without paying a dime', sold a package of lies by telling locals they'd be 'better off' with jobs at the plant which never materialised and misappropriated public funds. As for the Camorra, all three eco-docs cite instances and provide images of their callous disregard for the environment. The most striking footage, however, comes from *Biùtiful cauntri* and the undercover police surveillance through their wire-tapping and filming the Camorra at work. Here is one exchange, beginning with the Camorra referring to their illegal dumping as:

> the Berlusconi dump, yeah, first the plastic and tyres, then, after the elections (*which they know will see Berlusconi elected as PM*), the other (*toxic*) stuff … But keep it quiet, we don't want the communists (*referring to the local protestors, Del Giudice and the Legambiente ecology association*) getting wind of it. They've (*the Prefects, mayors, etc.*) agreed, we have made an agreement to avoid trouble with the locals.
>
> (*my italicised parentheses)*

It is not difficult to perceive from the above examples just how deep the networks of corruption are. As Deputy Public Prosecutor Donato Ceglie, who ran this undercover operation, makes clear:

> in the 1990s, for the first time, criminality infiltrated spheres of the judiciary, taking the form of an intricate relationship between parts of the business circles, deviated Masonic circles and in particular the Camorra which has infiltrated political circles and ended up corrupting the system of government Commissioners.

This triangulation of power – so clearly exposed by these eco-docs – begs the question, how do you prevent institutionally endorsed transgressions of the planetary and social boundaries especially if the judiciary then 'fails' to prosecute correctly in spite of the evidence? For such is the case here. At the closing moments of *Biùtiful cauntri*, we are told that a series of court cases were filed (in 2007) against, among others, several of FIBE's CEOs for numerous corrupt practices and the government Commissioner Antonio Bossolino for legal offences. However, in 2013, when these cases finally came to court, all were acquitted of any wrong-doing and most have continued in high office. Meanwhile, four out of ten farmers in Campania have lost their livelihoods. Incidences of cancer, birth malformations and rare tumours have arisen in the two critical areas: the land of pyres and the triangle of death – all of which was already attested to in the three eco-docs. Incinerators, when they are in place, merely add to this problem dispersing their fly-ash in a 2-km radius. The end titles of *Una montagna di balle* inform us that within two months of the opening of the incinerator in Acerra (in 2009), 'the presence of solid particles Pm10 in the Acerra air was 17 times superior to the health safety levels'.[15] The oncologist Antonio Giordano conducted a detailed study of the environmental relations between toxic dumping and cancer growth in Campania and found that there was a 12% rise in cancer and an increase in birth defects affecting the central nervous system by more than 80%.[16] Noting the source of this damage as an outcome of the tripartite apparatus of entrepreneurs, politicians and criminal groups, Giordano (writing in 2019) summed up this triangulation of waste, business and environmental destruction by coining the term *biocidio*:

> *Biocidio* (biocide) is the systematic violation of the right to health through poisoning of the environment and the consequential exposure of the population to pollutants. In Campania, it means a whole ecosystem has been sacrificed to narrow interests and greed by an apparatus of politicians, entrepreneurs and criminal groups that have turned the region into the 'biggest open-air laboratory of carcinogenesis in Europe'.[17]

Italy is not an isolated case – but it is certainly the one in Europe that has garnered the most vociferous protests over the longest of periods: 30 years and still ongoing, with no change to speak of from government or regional authorities. Here is what the 2017 European Parliament Report on waste management and transitioning to a circular economy in the EU has to say about Italy:

> The mechanical-biological treatment (MBT) of municipal waste has a relatively long history in Italy. It began around two decades ago, to currently reach more than 120 MBT plants across the country, which treat around 22.4 per cent of municipal solid waste (MSW). … Around 75 per cent of all crimes related to waste occur in Southern Italy, and are associated with the high share of landfills located in that region. The MBT of waste can *potentially* reduce the negative impact associated with the landfills and improve *political stability* within the country. … Despite the fact that Italians aim to improve the efficient use of resources, MBT alone has little or no impact on social values as the mixed collection of waste does not provide incentives to separate and reduce waste.[18]
>
> *(my italics)*

The report's answer to this ongoing dilemma? 'The transition to a more circular economy may require separate collection and citizen engagement to ensure both higher efficiency and prevention as well as higher quality'. You will note that no mention is made of the industrial dimension of waste production, the focus being solely on household waste which in Italy's case – at 8% of all waste and as our eco-docs have made clear – is not the primary problem the country faces. This gesture of waste-triage serves little purpose if it is not done at *all levels* of waste production, thereby bringing ecological stability. But, as we have already seen, the other levels are, for the most part, beyond control on both a legal and an illegal plane.

As *Clouds of Smoke*, *Trashed* and *Polluting Paradise* – eco-docs which provide similar narratives to those of the Italian trilogy – make equally clear, it is «the beyond control» that most fiercely needs addressing. Nor is this a case of local problems. This is a global predicament intrinsically tied into neoliberal economics. For that place of «beyond control» – what I've referred to as «deregulated madness» in *Film Ecology* (Hayward, 2020: 69) – is in the sector of money-production, the very sector that most easily escapes sanctions. Waste means money, the accumulation of which is dependent on the continuous flow of waste, whatever its provenance. Feed the incinerator beast that fills the coffers of the few, continue to store poisonous waste and all types of unrecycled rubbish in ecobales, or toss it into landfill sites and abandoned quarries, and, where necessary, set fire to garbage piles so there is room for another bumper bundle of rubbish; treat the oceans in the same exploitative way. All these modalities lead to wealth accumulation, to money-production that is institutionally legitimised by our century-long capital-growth economic model which in turn is endorsed by government, assisted by banks happy to finance the industrialisation of waste management and, finally, supported by current legal frameworks that have yet to see the value of legislating against practices of biocide and Ecocide.

In terms of Raworth's Doughnut we can see how the waste management practices described in this chapter are profoundly undemocratic. To that effect, 8 out of the 12 dimensions of the social foundation are transgressed, beginning with *networks* which, instead of abiding by a social contract (Raworth, 2017: 125), are established in such a way (the s/triangulation I spoke of) as to take advantage of a global crisis (waste) at the expense of both the planet's and humanity's *health* by imperilling two essential sources of sustenance, *water* and *food*. In bypassing the social contract, this triangulation of *networks* not only transgresses at least three of the planetary boundaries – *air pollution*, *chemical pollution* and *land conversion* – it also impacts directly upon a further four social foundation elements: *income and work*, *political voice* ('no one listens'), *peace and justice* ('those in power do nothing' other than militarise the sector and criminalise those who protest) and *social equity* (decisions are unilateral, handed down by those in power, locals are treated as ignorant of the 'facts', their illnesses dismissed as a consequence of poor diets). Figure 2.2 provides a summary visualisation mapped onto the Doughnut of this gloomy state of affairs.

This is a quite devastating visualisation that reveals the dangers of basing the purpose of waste on the ONE IDEA, ONE PRODUCT, ONE COMPANY, ONE DEAL discussed in Chapter 1. We can also see from the

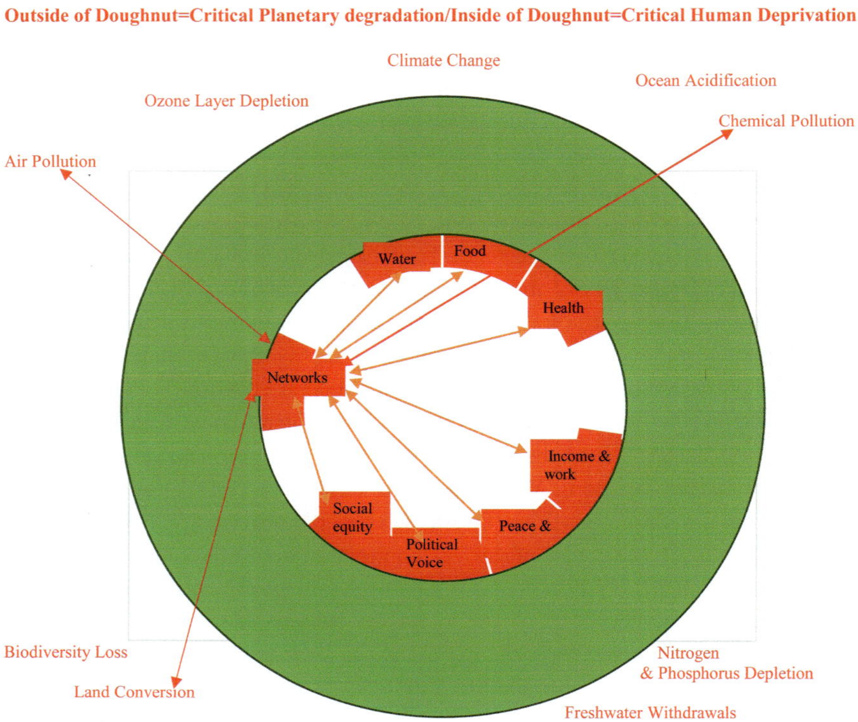

FIGURE 2.2 Mapping the pollution eco-docs' performative documentary practice and their visualising "the beyond control" – plotting the transgressions of the social and planetary boundaries by waste-management networks and the impact of their practices on local inhabitants.

diagram how *networks* are at the heart of this transgressive practice. The powerful network-triangulation – of economics (waste-merchants: businessmen and the Camorra)/local and national government/judiciary – is vividly represented, both in terms of its impact on the social dimensions (in double-edged orange arrows) and planetary boundaries (red arrows). Furthermore, one of the striking similarities with all the above eco-docs, in terms of their representation of waste management, is the way in which they embody Raworth's concept of the 'industrial model' which delivers 'strong profits' to business, but which 'is fundamentally flawed' because, in running counter to 'the living world which thrives by continuously recycling', it has 'broken the natural cycles apart' (212). Raworth calls this the 'economy of degenerative design', whereby we take=>make=>use=>lose. Figure 2.3 shows her diagram.

Garbage is the most visible example of the outcome of this practice of take and waste – central to which, as we see on the Doughnut diagram (Figure 2.2), is who owns the *networks*, for therein lies the power to do the damage. And in the above analyses we have shown how waste is the clearest symbol of over-consumption and

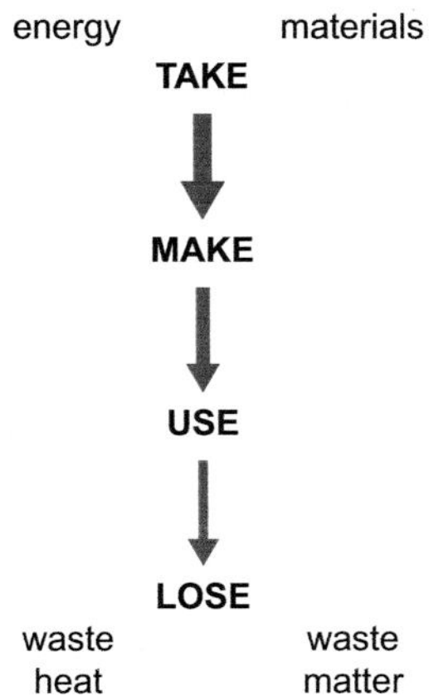

The caterpillar economy of degenerative industrial design.

FIGURE 2.3 Raworth's diagram of the degenerative linear economy (2017: 212).

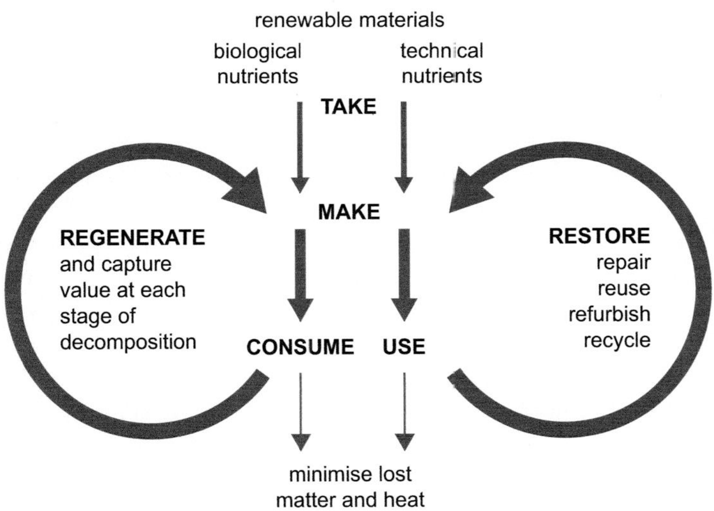

The butterfly economy: regenerative by design.

FIGURE 2.4 Raworth's diagram of the Ellen MacArthur Foundation's butterfly model for a circular economy (2017: 220).

economic inequality. But what if waste became its own regenerative economy? What if it were able to embody, instead, Raworth's butterfly diagram (see Figure 2.4)?

As a partial response to these 'what ifs', the final section of this chapter makes way for a detailed analysis of Lucy Walker's *Waste Land* (2010) which engages with these questions. So, let's take a look.

2.2 Trash as value, a very peculiar pecuniary circularity, the case of *Waste Land* and the Brazilian *Catadores*

Who would have thought that collectors of recyclable waste, known as *catadores*, scrabbling up and down the mounds of «household rubbish» dumped daily in the Jardim Gramacho landfill site outside Rio de Janeiro in Brazil, could both make a living wage through recycling and make art through that same process? This is the story (filmed over a period of two and a half years) that Lucy Walker's eco-doc *Waste Land* presents us with. By following the trajectory of the Brazilian artist Vik Muniz's interactions with the *catadores*, Walker unravels this extraordinary tale of the transformative value of making art from waste. The spine of the documentary is twofold: Walker's interest in trash and Muniz's desire to invite the *catadores* into his universe of art and thereby, as he puts it, 'change the life of people with the very material they deal with everyday'. This he plans to accomplish by entering into conversation with the *catadores*, hearing their stories, then enlisting a handful of *catadores* (seven in all) as both the subjects of and participants in the making of their own portraits by using garbage as artistic material. The proceeds from the sale of the portraits will then be returned to the *catadores* to invest in their cooperative organisation at all levels. We follow that process, but we also garner a great deal of knowledge about the *catadores*' way of life. And this is where we begin.

Let us first pause on the creation of the dump and its name – Jardim Gramacho. In 1978, the Military Dictatorship designated the district of Jardim Gramacho (30 km outside Rio de Janeiro) as the landfill site for the city's waste. Jardim Gramacho was an ecological wetland (of mangroves and silt) adjacent to Rio's Guanabara Bay. The name means garden of grasses (or marshland grasses), referring to the valuable ecosystem that was about to be destroyed (intriguingly, the name *Gramacho* has a possible unintentional secondary meaning: *grama* means grass and *cho* could be an abbreviation of the Brazilian word *chorume* which means leachate). By the time of the making of *Waste Land*, the 150 ha of the dump was 55 metres high and, because it stood on the unsteady marshland underneath, the sensation when walking about the dump was of 'walking on Jello' as Muniz puts it. The feeling of this instability is compounded by the dangerous nature of the recycling job itself, whereby *catadores* can easily get injured or suffer health issues from the escaped methane gases. On another scale of harm, there is the effect of the *chorume* leaching down into Guanabara Bay killing off the fish population.

By the mid-1990s, Jardim Gramacho had become one of the world's largest landfills and was in receipt of 70% of Rio's garbage from both wealthy and poor districts. Over the two decades since its opening, it had also become the unregulated wasteland of a *catador* squatter community living and working on the landfill. In 1996, Rio's sanitary department intervened to formalise the role of the *catador* which, in turn, led to considerable abuse of power by the officials and guards staffing the landfill – including deciding on a daily basis who they would let in to work and obliging the *catadores* to sell to middle-merchant scrap dealers, thereby reducing their income. In 2004, in response to this exploitation and to threats of closure, the *catadores* formed an association, the Association of Collectors at the Metropolitan landfill Jardim Gramacho (ACAMJG), under the leadership of Tiao Dos Santos. More than half of the landfill's 3000 *catadores* joined up, the rest choosing to remain independent. By 2006, the ACAMJG was functioning as an autonomous economic entity with a decentralised system of recycling, whereby *catadores*, equipped with their own trash barrel (at a cost of $30), collect recyclables (women tending towards plastic and the men toward the heavier elements) to then sell on. The ACAMJG deducts 15% of their earnings for running costs such as their own transportation system. So as to protect its members' earnings, the ACAMJG bought two trucks in order to bypass the intermediary scrap dealers. One truck brings in pre-sorted paper and cardboard donated from businesses; the other picks up all the recyclables. These then get sold directly to recycling companies, which allows the materials to be sold at double the normal price. By 2007, ACAMJG obtained funding and built a recycling plant (*polo*). In 2010, thanks to the proceeds from the sale of the *Waste Land* portraits, a medical clinic as well as a day-care and skills-training centre were set up. That same year, as a result of the combined effect of *Waste Land* being nominated for an Oscar, the portraits' exhibition at Rio's Museum of Modern Art and their subsequent sale, the eco-doc furthered its impact value. This foregrounding of the value of the recyclers' work strengthened the hand of the ACAMJG, and the Brazilian government decided to enact into law the professional recognition of the *catadores* (who number over 800,000).[19]

This is the ecosystem *Waste Land* brings to us. We also learn that, thanks to the rich pickings, the *catadores* over the years retrieved some 7000 books for their own library, overseen by Zumbi – one of the *catadores* interviewed and selected for a portrait. A source of learning for all who wish to access it, this library collection enabled Tiao to become self-taught. Amongst these books, Machiavelli's *The Prince* was life-changing, as he explains:

> Without this, I would not be where I am today. The day I decided to fight for our rights, I believed everyone would support us. But no, what I discovered was there are a lot of people with vested interests in keeping us stuck in poverty. Machiavelli talks a lot about power and the games of power. So I began to adapt his texts to my situation and I learnt that I needed to be less aggressive and more strategic.

To further this knowledge-transfer, Tiao took courses in Brazilian history, political organisation, social movements and civil rights, all of which led him to set up the foundation, with four others, of the ACAMJG.

In the various interviews conducted throughout *Waste Land*, a constant echo is the *catadores*' statement to the effect that their work at the landfill is an 'honest way to earn a living', that it 'is a much better alternative than turning to crime, prostitution and drugs' (the only other viable ways for the poor of accessing money). On average they expect to make $20–25 a day (by way of comparison, in 2010, the minimum wage in Brazil was $180pm, and the average wage around $450pm). Mothers explain how they are saving up to put their own children through education and towards a better life. Some *catadores* have worked the landfill most of their lives, and many speak of the pride they take in their job. We quickly gain a feeling for their sense of community as the camera goes around their work-place and living spaces at the foot of the landfill – shanty huts they may be but, as they explain, at $8 a week, they serve the purpose of shelter (and the *polo* has fully functioning latrines and showers). Many of the workers have family elsewhere to whom they return from time to time before needing to come back to work. Irma, another of the portraited *catadores* and the on-site cook (providing nurturance to the workers from reclaimed food), joyfully exclaims how happy she is to be there, adding 'I don't let anyone go hungry'.

These testimonies attest to a self-awareness and self-value – something, as the *catadores* themselves recognise, that is not accorded them by the 'classism' which reigns in Brazil wherein they are seen as the lowest of the low. One of the women, Isis, makes this point: 'when people know what I do for a living', she says, 'they treat me as if I'm dirty'. Even Muniz, who had himself grown up suffering the indignities of poverty, was surprised by what he found: 'these people are the other end of consumer culture, I was expecting to see people who were beaten and broken'. However, when he adds 'but they are survivors', it is here that we can see how his perception of the *catadores* at the beginning of his project is skewed by his own memories of lack which he had projected onto the workers' own form of living – and it is this that makes the eco-doc so fascinating. Kathleen M. Millar (2018: 9) – whose coining of 'form of living' I have found so clarifying – puts her finger on the source of this misconception when she explains (in her study of the Jardim Gramacho *catadores*) that 'one of the problems with seeing the work of the *catadores* as a survival strategy or a last resort is that it reduces their existential concerns to the merely economic' (11). As Muniz himself had done – but, as a result of spending time with the *catadores*, he came to a different understanding. For, much as Muniz is committed through his art to help socially deprived communities and, in this particular project, is engaged in bringing his own art form of living and sharing it with the *catadores* to help transform their lives, so too, over the two and half years of working with them, the *catadores* enlightened him as to a different mode of living life, so much so that he declared he found it 'difficult to return to everyday life'. As Millar explains, rather than 'survival', the *catadores*' forms of living constitute a divergence 'from state-regulated officially recognised institutionally organised forms of work in capitalist societies' (13). The *catadores* do not fit within the capitalist model of wage labour which, as Millar indicates, is upheld as a source of social belonging even though it simultaneously creates a manageable and industrious population (9). In short, the wage-labour nexus functions both as a form of social control and respectable-belongingness. Indeed, by contrast, since officially the *catadores* are unemployed, their collecting is not seen as real work, but rather as begging

(10). Millar continues: 'as unemployed workers who sift through garbage on a city dump, *catadores* seem to exemplify in extreme form today's notion of a disposable life' (8). They too – in capitalist terms – are on the garbage heap, broken and discarded, which is what Muniz had expected to find. What he discovered, instead, was an entire ecosystem with, at its core, waste as its economic facilitator. As Millar astutely remarks, 'if *catadores* are superfluous to capital accumulation, then it becomes impossible to ask how the materials they collect are tied to a 200-billion-dollar global recycling industry' (8)!

Tiao's insistence that they are not garbage pickers but 'pickers of recyclable materials' underlines this refusal of the *catadores* to accept identification through metaphors of waste (which include 'superfluous', 'beggars'). As Lucy Walker asserts, the work executed by these workers – the effect of sifting, selecting and separating recyclables – means 'there's a real role for these people who are tremendous environmental stewards and really know their stuff regarding garbage and recyclables'.[20] Muniz points out that their careful work means that half of household waste is recycled and does not go out into the rivers and pollute – so they are part of the containment of the damaging effects of human over-consumption. But, in that it feeds both their alimentary and financial needs, that saved part of waste also goes on to source and resource the *catadores*. As we have seen, Irma's picking over the rejected food-produce to create bodily nourishment is as much valued as are the other outcomes of recycling which include materials for dollars, but also waste as food for thought, such as books.

It is into that same materiality that Muniz and the *catadores* delve to both source and resource the portraits. Source, because all types of selected objects, including bath-tubs, toilet seats, fabric and other paraphernalia, are used to compose the mise-en-scène of the seven portrait photographs that Muniz first takes and then prints onto huge canvases (approximately, 7 m by 3 m). Resource, in that, assisted by the *catadores*, these super-sized images are subsequently 'photoshopped' by placing smaller elements, such as the caps off cola bottles, plastic tabs, nuts and bolts, to highlight and shadow the faces and bodies of the *catadores*' portraits before the second photograph process. Waste is now transformed into yet another form of nurturance – art – the responsibility for which falls, in this instance, to Muniz who both selects the kind of portrait he is aiming for and the required pose. The *catadores* have shared their knowledge with him. Now it is his turn – he explains his choices to his subjects and, as we have already explained, involves them in the actual making of the portraits. Of the seven portraits, four directly reference 'grand masters'; the other three, cultural prototypes – all are germane to the individual chosen. Tiao, the politicised leader, is selected to embody David's *Death of Marat*; Zumbi, the book retriever and library guardian and thereby dispenser of knowledge, is Jean-François Millet's *The Sower*; Carlao becomes the emblem of endurance as Barbieri's *Atlas Holding the World*; Isis meanwhile poses as Picasso's *Woman Ironing* (with perhaps an intentional reference to her cleanliness). Irma is emblematic of the provider in her portrait as *The Bearer* in which she proudly transports a huge basket of garbage on her head; Suellen, a young mother whose children live with her mother elsewhere and for whom she works hard for a better life, is adorned as *The Madonna* (with two children); and finally, Magna, a free-spirited woman full of humour, who comes and goes from the landfill, is portrayed as *The Gypsy*.

What a distance the garden marshlands of Jardim Gramacho have travelled since 1978, from desecrated land to provender of aesthetics! What a defiant outcome, to sell back to the rich *their* garbage in the form of these seven portraits (to the tune of $250,000) and moreover to sell *back* to the rich the recycling *skills* of the poor. It certainly puts an interesting twist on labour-value, since no exploitation occurs in this contract of exchange-by-return. But what *Waste Land* most significantly shows us through the *catadores*' form of living is an economic model based in regenerative practices from which they gain so much more than just money. Theirs are productive lives, despite the fact they are outside the wage-labour nexus of neoliberalism from which their status as 'Class E' (the poorest section of society) more or less excludes them. Despite their poverty, these individuals see the worth of their work, which consists in removing materials that are recyclable, up to 50% of household waste – enough to have impact value for the environment and (of all ironies) contribute to a '200-billion-dollar global recycling industry' (Millar, 8).

Waste Land's *catadores* provide us with a living response to Raworth's call to 'create to regenerate', to practice in terms of 'waste means food', be it 'food scraps or scrap material' which can become 'the source materials for the next' materials (221). If we refer back to Raworth's diagram of the Ellen MacArthur Foundation's butterfly model for a circular economy (Figure 2.4) we can see how the *catadores*' forms of living precisely embody this practice which Raworth describes as follows:

> The key to making this work is to think of all materials as belonging to one of two nutrient cycles: *biological* nutrients such as soil, plants and animals, and *technical* nutrients such as plastics, synthetics and metals. The two cycles become the butterfly's two wings, in which materials are never 'used up' and thrown away but are used again and again and again through cycles of reuse and renewal.
>
> *(221)*

Waste Land documents the *catadores*' model of 'taking and making' as they source and resource from both the landfill dump itself (as a space for living) and the dumped garbage (as a means for living) in a diverse set of ways and activities. Most vitally, too, as Millar explains, these diversified forms of living 'allow us to consider how work as a category of action may be constituted by a wide range of practices beyond the purely economic' which defy the 'work/life and labor/leisure dichotomies' that dominate the wage-labour model (10). The *catadores*' form of living becomes, thereby, both a 'livelihood and a way of life' (11). There is socialising, playing games, 'listening to the radio, lounging around', 'cooking and sharing meals' and recycling (ibid). What we also witness in *Waste Land* is a community committed to a social contract form of living based in an economic model of association (the ACAMJG) and shareholding (agreed percentage of investment in infrastructure), a social contract that provides security and flexibility (*catadores* come and go and return, invest, when there, through their recycling).

In terms of Raworth's Doughnut and the 12 social foundations, we see how the *catadores* attain at least 8 of the elements that place them in a safe and just place for humanity. By establishing their own *network* (ACAMJG), they resist dominant networks that seek to exploit their labour. By organising themselves, they not only gain a *political voice* and ensure a relative level of *peace and justice*, they also create an environment where *social equity* prevails (equal investment in ACAMJG, equal access to *income and work*, equal access to up-skilling). Their labour works to secure their children's *education*, and their resourcefulness has led some to self-educate from their recycling efforts. *Gender equality* is facilitated by the day-care centre in that mothers can freely work on the landfill. Finally, no one goes hungry (*food*).

This is not to say that the *conditions* under which these forms of living are experienced – in poverty – are how things should be, nor that any level of poverty is acceptable. However, as things currently stand, the *catadores* – who are overwhelmingly African-Brazilian – form part of Brazil's poorest economic group ('Class E') which, according to the 2010 Brazil census, is made up of 76.3% African-Brazilian and 23.7% white.[21] Considering that over half of Brazil's population is African-Brazilian, of whom such a high percentage live in poverty and without adequate education, it becomes evident how limited options are for a great many.[22] Viewed in that light, recycling becomes a viable economic system for the *catadores* both to protect themselves from the other predominant routes of income-generation (drugs, prostitution or theft) and to assist their children to a better life out of indigence. And that is the story *Waste Land* has presented us with, along with Muniz's desire to help bring about change, through collaborating with his art.

In writing the above, I am not blind to the criticisms levelled at Muniz's intervention, that it prompts ethical questions (ranging from exoticisation of the precariat, to whether any lasting changes are possible) and that by elevating the *catadores*' 'images into luxury objects', this 'revaluation ultimately endorses and reifies the capitalist logic by which humans are reduced to their labor, instrumentalised to produce value for capital'.[23] But these views seem to miss the point that the *catadores* have charge of their own economies, their forms of living. We recall, also, that at the same time as the *catadores* work outside of the neoliberal economic model, nonetheless, the net effect of their work impacts upon the global recycling industry by virtue of the contribution of their work skills. Furthermore, the relationship between the *catadores* and Muniz was one of shared experiences. Transformative exchanges (much like the recycling) went both ways. The *catadores* brought Muniz into their everyday narratives of lives lived and brought him new insights into his earlier life which he, too, experienced on the margins. Their inclusivity formed a web of humanity that profoundly affected him: 'magical things' happening right in front of him, as he puts it. For his part, Muniz did make a valuable contribution to the well-being of the community he worked with and allowed for changes to take place, some of which I have already mentioned. In terms of 'exoticisation', the answer comes from the *catadores* themselves when they attended the exhibition's opening. Irma exclaims 'I became famous all over the world in garbage'; Magna tells Muniz 'you have no idea what this means to me'; Isis, who spoke of the social stigma attached to her work, recognises the impact of Muniz's portraits when she declares 'sometimes we see ourselves as so small, but people out there see us as so big, so beautiful'. The exchange value of Muniz's work with them and theirs with him was one of a shared dignity, a mutual recognition of their humanity.

Interestingly, Jardim Gramacho was closed in 2012, two years after this film's release. Given that Rio was hosting, in turn, the COP 21 (2012), the World Cup (2014) and the Olympic Games (2016), the government felt obliged (at last) to sanitise the area. About half of the *catadores* received compensation for loss of work; several, thanks to Muniz's donation were able to find different forms of work; still others moved to new recycling plants to work in safer conditions thanks to contracts secured by ACAMJG.

In terms of the significance I attach to *Waste Land* and its relationship to Raworth's Doughnut Economics, it is of course the regenerative design of the *catadores*' model that I've chosen to highlight. I do so because I read it as metonymic of what can be done – it is surely extraordinary that the *catadores* on a daily basis take out 200 tons of recyclable material from the landfill (the film's Brazilian title *Lixo extraordinario*, which translates as *Extraordinary* or *Spectacular Trash*, is most apt). This represents half of the estimated *recyclable* household material. Relative to the overall daily dumping of 7000–9000 tons of trash, this skilled work clears up less than 1% of the city's garbage. The Jardim Gramacho wasteland remains, therefore, an ecosystemic disaster, much as we saw in the Italian trilogy. And, as the other eco-docs mentioned in this chapter attest, it is a wide-spread phenomenon all over the world. Brazil relies heavily on landfills (some 6000) to «manage» its waste deposits which are, as we've noted, completely unprocessed.

Thus, this tiny (in terms of volume) example of regenerative recycling provided by the Brazilian *catadores* can only act as a guide in principle as to what needs to be done in relation to our consumption habits and the processing of our household waste where renewables and non-renewables are concerned. It also functions an indicator towards the larger question of all waste products. If recycling is to be made to work, then, first, there must be less "throw-away" and, second, clean methods of processing recyclables. Incinerators are not the answer for two major reasons. As we learnt from *Una montagna di balle*, the cost of developing and running them means that they must be constantly fed with waste and there is no necessary assurance of the quality of the waste being processed. The characteristics of incineration emissions, fly-ash and residual ash are affected by the types of waste fed into the incinerator, its combustion efficiency and the degree of emission control of that facility. Whilst filters serve to control the minuscule airborne particles, nonetheless, a percentage enters into the atmosphere. Moreover the filters are so expensive they do not get replaced with the appropriate regularity (as *Trashed* pointed out).

Nor are Clean Development Mechanism projects (CDM), which reduce and recycle methane emissions from landfills, the answer. This was a route Brazil followed (beginning in 2004–5) and, in the light of the 2012 COP21, considerably expanded upon – increasing its CDMs by 30% to 413 projects. However, the CDMs are really an offset scheme (developed under the Kyoto Protocol of 1997) which allows rich countries to meet some of their emission-reduction targets by investing in green projects in developing countries and buying their emission reductions to offset their own excesses. For every tonne of *avoided* greenhouse gas, a carbon credit (CER) is issued. But of course this is not a solution to either waste or carbon emissions. Rich countries buy themselves out of their difficulties (greenwashing), and in return Brazil (in this instance) – with 39% of the CDM business in Latin America – becomes a rich partner in the scheme. Meanwhile, very little control is exercised over the real figures involved in the number of CERs exchanged.[24] Instead of waste disposal we should be thinking in terms of a circular economy of recycling through repair and re-use. It is estimated that such an economy would generate new jobs from 3- to 12-fold within the recycling industry.[25]

As for non-recyclables, most of which incidentally are by-products of non-renewable resources, it is evident that a progressive diminution of the most damaging materials must occur. The three most critical ones are not only the chemical or synthetic manipulations of natural, planetary elements, they are also those most significantly linked to war, be it nuclear waste (uranium), plastic products of all kinds (oil) or chemically derived synthetic fertilisers, insecticides and pesticides (nitrogen). The impact of the nuclear and oil industries on the biosphere in terms of their transgressions of the planetary and social boundaries was discussed in Chapter 1. Some of the knock-on effects have been at the core of this current chapter as indeed has humankind's individual role in the polluting of planet Earth. Now, it is time to turn our attention, in Chapter 3, to the synthetic world of the agribusiness industry – and those resisting it.

Notes

1 See: <https://www.bloomberg.com/news/features/2019-07-11/how-the-world-can-solve-its-2-billion-ton-trash-problem> (accessed 31 December 2020).
2 Candida Brady cited in: <https://en.wikipedia.org/wiki/Trashed_(film)> (accessed 10 June 2020).

3 I include this eco-doc, even though it was subsequently withdrawn after a brief airing on YouTube. Monty-Pythonesque in tone, this short film was deemed too controversial in its advocating 'blowing people up' who refused to be enthusiastic about the 10:10 campaign to reduce CO_2 emissions.
4 Fatmir Terziu writing about the making of *Clouds of Smoke* <http://fatmir-terziu.blogspot.com/2007/06/making-clouds-of-smoke.html> (accessed 10 June 2020).
5 Press release for *Polluting Paradise* <https://www.the-match-factory.com/catalogue/films/polluting-paradise.html> (accessed 31 December 2020).
6 Report published by European Union Action to Fight Environmental Crime <https://efface.eu/sites/default/files/EFFACE%20Victims%20in%20the%20Land%20of%20Fires_0.pdf> (accessed 28 December 2020).
7 Angelone's approach is to examine how the documentary practice in this trilogy functions performatively to give voice to the victims of Italy's waste management. Angelone, A. (2011) 'Talking Trash: Documentaries and Italy's "Garbage Emergency"', in: *Studies in Documentary Film*, 5:2 and 3, 145–56. Another valuable study on eco-docs and pollution is Karl Schoonover's 'Ecocinema and the Toxic', in: *European Journal of Media Studies* <https://mediarep.org/bitstream/handle/doc/3285/NECSUS_2_2_2013_483-507_Schoonover_Documentaries_without_documents.pdf?sequence=1&isAllowed=y> (accessed 18 January 2021).
8 See: <https://efface.eu/sites/default/files/EFFACE%20Victims%20in%20the%20Land%20of%20Fires_0.pdf> (accessed 4 January 2021).
9 There are some 1300 abandoned quarries in Campania, including privately owned ones. But only a handful of these have been obtained for legal use.
10 In fact, the first one was not inaugurated until 2009 (by Berlusconi), a year after FIBE lost its contract.
11 A third of Italy's waste disappears in this manner and Campania is where most of it turns up. By 2004, this illegal business amounted to the equivalent of 10% of Italy's GDP! One assumes that figure is even higher now, since there has been no abatement in illegal dumping. See: <https://www.theguardian.com/world/2004/oct/14/italy.sophiearie> (accessed 28 December 2020). For the health effects of this see: <https://pubmed.ncbi.nlm.nih.gov/29320415/> (accessed 9 January 2021).
12 See: <http://volcanology.geol.ucsb.edu/soil.htm> (accessed 12 January 2021).
13 The non-urban area of Campania is 12,420 square km of which, according to Del Giudice, some 1765 square km has been polluted 'as a result of legal and illegal waste-dumping'. According to other sources, it is more like 4000 square km that are polluted (see: <http://www.ejolt.org/2013/11/waste-conflicts-in-campania-and-the-common-struggle-against-biocide-join-hands/> (accessed 12 January 2021)). *Legambiente* (the Italian environmental NGO) estimates that, in the last 22 years, over 410,000 trucks coming from northern Italy and from abroad have illegally disposed of some ten million tonnes of toxic waste in the region, mainly residues from medical products, paints, used tyres, dioxin and allegedly even nuclear waste (see: <https://www.theguardian.com/cities/2016/oct/11/hell-earth-great-urban-scandal-life-rubbish-dump> (accessed 18 December 2021)).
14 In the period relating to these eco-docs (2000–9) alone, reprimands and fines have occurred four times.
15 For the full story on the Acerra A2A incinerator, opened in 2009, see: <https://ejatlas.org/conflict/urban-waste-incinerator-of-acerra>. It appears it has since been taken over by Fisia Italimpianti SpA – a new group 'born' from Salini Impreglio (so in a way back to the former contracted FIBE!). Fisia proudly displays this monstrous construction on its webpage and describes it as follows: 'part of the integrated Waste Management Campania System, Acerra WtE plant is one of the largest in Europe, being capable to dispose 600,000 tons/year of selected MSW, and generates 600 million kilowatt-hours per year of electric power, a quantity enough to satisfy the needs of 200,000 families'. See: <https://www.fisiait.com/en/business-areas/sustainable

-waste-management/waste-to-energy-plant-acerra.html> (both accessed 7 May 2021).

16 See: <https://en.wikipedia.org/wiki/Antonio_Giordano> and see also <https://www.news-medical.net/news/20191216/Cancer-patients-in-Southern-Italy-have-higher-concentration-of-toxic-metals-in-their-blood.aspx> (both accessed 14 January 2021).

17 See Sonia Goicoechea's article in: <http://www.ejolt.org/2013/11/waste-conflicts-in-campania-and-the-common-struggle-against-biocide-join-hands/> (accessed 14 January 2021).

18 See: <https://www.europarl.europa.eu/RegData/etudes/STUD/2017/581913/EPRS_STU%282017%29581913_EN.pdf> pages 90–1 (accessed 16 January 2021).

19 To give an idea of the importance of the *catadores*' work to the local community, around 50% of the 40,000 local inhabitants of the Jardim Gramacho district depend directly or indirectly on the economic activity emanating from garbage collection. See: <https://www.maxwell.vrac.puc-rio.br/13053/13053_4.PDF> (accessed 31 January 2021).

20 See: <https://www.npr.org/2011/02/21/133941168/Trash-Pickers-Turn-Rio-Waste-Land-Into-High-Art?t=1611675656507> (accessed 26 January 2021).

21 For these figures see: <https://www.theguardian.com/world/2011/nov/17/brazil-census-african-brazilians-majority> (accessed 22 January 2021). Brazil has a five-class system (A–E). Class E refers to people who earn the minimum wage, such as cleaners, street sweepers and also unemployed people. See: <https://thebrazilbusiness.com/article/social-classes-in-brazil-1453802521> (accessed 19 April 2021).

22 According to a recent report on education in Brazil, poverty affects 70% of African-Brazilians where illiteracy runs high. See: <https://sites.miis.edu/educationinbrazil/diversity-and-access-to-education/>. Within that 70% there is the category 'living under the poverty line', which affects a third of those African-Brazilians living in poverty, with a further 9% living in 'extreme poverty'. See: <https://agenciabrasil.ebc.com.br/en/economia/noticia/2019-11/blacks-make-majority-brazils-unemployed-and-informal-workers>. I should add that there is no official poverty line applied in Brazil which makes quantifying those living in poverty quite difficult. Yet another study <http://ftp.iza.org/dp2809.pdf> finds also that one-third of African-Brazilians do indeed live *below* the poverty line. Thus, I have been guided by these three studies (all accessed 19 April 2021).

23 This less favourable take on Muniz's project is from Christopher Schmidt's thought-provoking essay 'Vik Muniz's Pictures of Garbage and the Aesthetics of Poverty'. See: <https://www.researchgate.net/publication/321173530_Vik_Muniz%27s_Pictures_of_Garbage_and_the_Aesthetics_of_Poverty> (accessed 22 January 2021).

24 See this study of Brazil's CDMs in: <http://www.scielo.br/scielo.php?script=sci_arttext&pid=S1414-753X2016000300199> (accessed 29 January 2021).

25 There are many interesting statistics on current waste patterns and possible solutions supplied by a recent 2017 EU report *Towards a Circular Economy — Waste Management in the EU*. I refer you in particular to pages 28–31; 78–82. See: <https://www.europarl.europa.eu/RegData/etudes/STUD/2017/581913/EPRS_STU%282017%29581913_EN.pdf> (accessed 16 January 2021).

3

BAD-HUSBANDRY VERSUS EARTH-AS-PROVENDER, A TALE OF TWO CULTURES

Agribusiness' monocultural-monopolies and Agroecology's pluriculturalism

Two stories are told by the eco-docs in this chapter on Agribusiness and Agroecology. The first is a story of *Accumulation of Wealth by Dispossession*[1] and the second is about a *Regenerative System of Growth for the Common-Wealth*. The former starves the Earth, the latter nurtures it. There are 11 eco-docs in the first domain: Agribusiness, which includes monocultures, agrochemical products such as pesticides, herbicides, fungicides and fertilisers and bio-technology such as genetically modified seeds (GMOs). In the second, Agroecology, there are ten eco-docs – five of which look at the damage caused by agribusiness practices and offer a panoply of workable solutions that have been implemented world-wide, thus proving that there are alternative methods which, whilst taking into account what the Earth can carry, are able to produce enough to feed the world (they are *Demain* and its follow-up *Après demain*; *Au nom de la terre*; *Solutions locales pour un désordre global*; *Cultures en transition/ Voices of Transition*). Of the other five in this category, two look at small-holders' practices (market-gardening in *Anaïs s'en va-t-en guerre*, and peasant farmers' traditional ways *La Vie moderne*), two at reforestation and tropical forest protection (*Forests of the 21st Century* and *Forgotten District* respectively) and the fifth (*Just Do It*) is about ecological-activists' protest tactics.

Table 3.1 provides a list of eco-docs.

In this chapter, I propose to look at these two stories in turn, but rather than examining each eco-doc as an individual case, I think a better approach would be to look at representative clusters. Thus, the first section, on Agribusiness eco-docs, will focus primarily on chemical and bio-technology as it affects the land, the insect world, in particular bees, and farmers caught up in this so-called Green Revolution.[2] Also known as the Third Agricultural Revolution, this Green Revolution began after World War Two and is still ongoing. It is based in technology-transfer derived from war technology (chemical warfare products transformed

DOI: 10.4324/9781003130246-3

TABLE 3.1 Table of Agribusiness and Agroecology eco-docs

I. Agribusiness: bio-diversity loss, land or space conversion, agrochemical products, monocultures and concentrated food production	*II. Agroecology and pluricultural solutions*
1. *Darwin's Nightmare* Austrian/Fr/Bel (2004)	12. *Anaïs s'en va-t-en guerre/Anaïs Goes to War* Fr (2014)
2. *Earth* UK/Ger/US (2007)	13. *Après demain/After tomorrow* Fr (2018)
3. *Les Enfants nous accuseront/Our Children Will Accuse Us* Fr (2008)	14. *Au nom de la terre/For the Sake of the Earth* Fr (2013)
4. *Humain/Human* Fr (2016)	15. *Cultures en transition/Voices of Transition* Fr/Ger (2102)
5. *Idiot Cycle* Fr/Can (2009)	16. *Demain* Fr (2015)
6. *McLibel* UK (2005)	17. *Forests of the 21st Century* UK (2009)
7. *Le Monde selon Monsanto/The World According to Monsanto* Fr (2008)	18. *Forgotten District* Fr/UK/Ger (2008)
8. *More than Honey* Switzerland (2012)	19. *Just Do It* UK (2011)
9. *Roundup face à ses juges/Round-up Facing Its Judges* Fr (2017)	20. *Solutions locales pour un désordre global/Think Global, Act Local* Fr (2010)
10. *Le Sel de la terre/The Salt of the Earth* Fr (2014)	21. *La Vie moderne/Modern Life* Fr (2008)
11. *Vanishing of the Bees* UK (2009)	

into pesticides, insecticides, herbicides and fertilisers) and the development of new breeds of high-yield crops intended to increase global food production. The second section, Agroecology, will concentrate on five eco-docs, selected because they provide a broad range of regenerative solutions to food production.

But first, because these points recur frequently in our eco-docs, here are some background facts as to how agriculture currently functions world-wide (see also Figure 3.1):

Between 15 and 18% of the global surface is considered habitable land.

Of that surface area 50% is used for agriculture.

Of that 50%, 77% goes on livestock (including feed for livestock); 4% on biofuels (mostly to produce ethanol to blend with petrol) and 19% on crops production (food we humans eat).[3] And as we shall see, below, 10% of this 50% of arable land is currently used for GMO crops.

You will not be surprised to learn that 70% of agricultural land pertains to large farms (over 50ha), and a further 12% to medium to large farms (from 5–50ha).[4] This means, overall, that 82% of arable land belongs to this sector of medium to large farms.

Small farms (those with up to 2 hectares of land) represent 12% of the total agricultural area, other small farms (2–5ha) 6% and, together produce between 36% of our food supply; (the other 64% or so of our food supply coming therefore from the medium to large farms).[5]

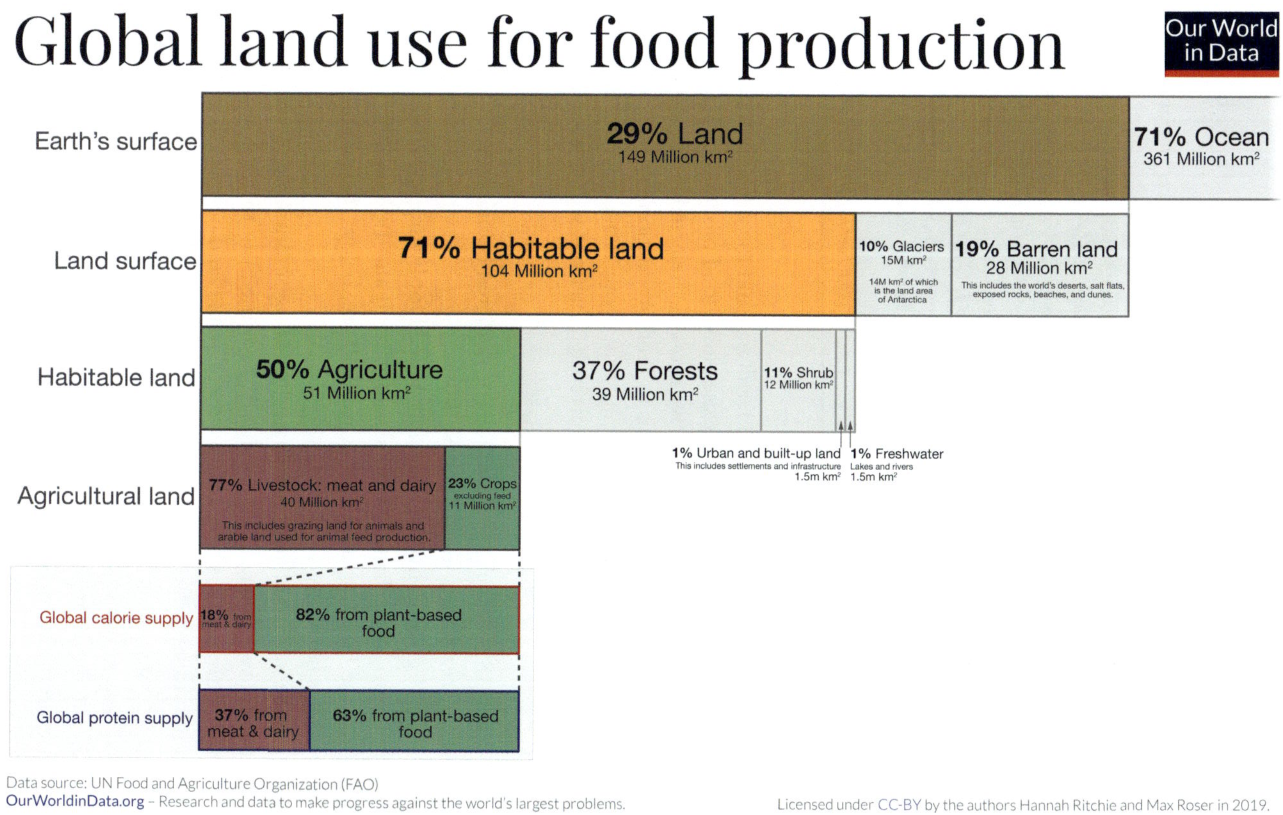

FIGURE 3.1 UN Food and Agricultural data – global land use.

In terms of the cause of world-wide land degradation and its link to agriculture practices the figures are as follows:

Overgrazing accounts for 35%
Deforestation, for the purposes of plant crops (soya, palm oil, maize), 30%
Overcultivation, especially of monocultures, 28%.

As you will note, all three of these are almost entirely attributable to large farm practices.[6]

Lastly, in terms of world-wide water consumption:

Agriculture accounts for 70%
Industry 22%
Residential 8%.

Overall the USA is the greatest consumer of water (on all three fronts).[7]

And what about the other 50% of habitable land? Well, first of all, forests make up 37%, the other 13% constituting a mixture shrub land (11%), freshwater lakes and rivers (1%) and urban living (1%).[8] Figure 3.1 is a clarifying Our World in Data chart. [9]

Since the 1970s there have been huge demographic shifts both in terms of human migration to the urban environment and population growth (3.5 billion to 7.8 billion in 2020).[10] Currently, in 2020, approximate figures suggest that between 24 and 32% of the world's population lives on the land (rural living), 68–76% in the urban environment (this breaks down into 85% of the industrial world's population living in urban environments, versus 35% of populations in developing countries).[11] In terms of proximity to flood risk, 10% of the world's population lives 10 metres above sea level, and 40% within 60 miles (100 km) of the coast. With rising sea levels – thanks to the melting of both arctic ice caps and glaciers around the world, as the eco-doc *Earth* so readily exemplifies – a significant percentage of the world population's well-being is clearly in danger. Of further concern still, given that a mere 1% of the world's habitable land is designated as urban, with just over half of the world's population (4 billion) occupying that environment, it is inevitable that serious problems do and will continue to arise (not enough habitable spaces, overloaded infrastructures, pollution, health issues).[12]

From the above figures on agricultural use, it is equally clear we need to change our food consumption practices away from a dependency on meat. This would reduce both livestock numbers and overproduction of animal feed crops (80% of which are GMO products[13]), thereby reducing also the dependency on herbicides, pesticides and fertilisers and water consumption. We get ample evidence in the Agroecology eco-docs that this transformation of agricultural practices is achievable and that we do have the ability to make shifts in our thinking and patterns of behaviour, on both a small and large scale. Let us now explore these issues through the eco-docs before us.

3.1 Accumulation of wealth by dispossession

3.1.1 Agribusiness in the dock – Part One: Le Monde selon Monsanto/The World According to Monsanto *(2008) and* Roundup face à ses juges/Round-up Facing Its Judges *(2017)*

Nine years separate these two eco-docs by Marie-Monique Robin. Both films reveal Robin's commitment to rigorous research, plus a willingness to show her own documentary process. Robin works by 'chapters' (I use quotation marks here because they are not necessarily numbered in the film). Each one is delved into in depth. In the case of *The World According to Monsanto* (the international release title), there are ten chapters, each one prompted by a Google search, the first one being 'Monsanto' itself. This initial enquiry then makes way for a preparatory investigation on-line, followed by telephone interviews with agencies (governmental, multinational corporations and so on) and other sources such as agriculturalists (large and small). Lastly, there is on-the-ground documentary filming, accessing people's testimonies (from Europe, Canada, India, Latin America and the USA). The structuring element, ultimately, however, comes down to the four elements associated with Monsanto. These Robin examines in great detail, exposing the hold this multinational has over the global food system and thereby economy. These four elements are PCB coolant fluids, bovine hormones, Roundup and GMOs. The findings of this documentary make plain that far from providing a Green Revolution (as Monsanto claims and governmental and international agencies have asserted), it is a case of a Green Desert that Monsanto has left in its wake, as one Paraguayan farmer puts it.

Eight years later, this tale of ecological harm enters the realm of the judiciary. *Round-up Facing Its Judges* (the international release title) documents (in six chapters) the procedures taken by the International Monsanto Tribunal (IMT, composed of five internationally renowned, now retired, judges) at the International Criminal Court (ICC) in the Hague. The ICC has as its foundation and governing document the Rome Statute (entered into force in 2002) and adjudicates on the four crimes against peace: crimes against humanity, genocide, war and aggression. Tribunals, however, are not a court, and can only make recommendations to the judiciary of the ICC. Robin's eco-doc follows the IMT's hearing of the evidence brought against Monsanto. The multinational was invited to attend and defend their case, but declined. Testimonies from 28 witnesses were heard around four key human rights (as established in the UN Declaration on the Right to Development, 1986): the right to a healthy environment, the right to food, the right to health and the right to information and independent scientific research. The IMT concluded that Monsanto did not act in conformity with basic human and environmental rights. The IMT found for the case that international law needs to change to incorporate the protection of the planet as a human right which would allow climate criminals to be brought to court and victims to gain redress. To this end the IMT recommended to the ICC that Ecocide should become the fifth crime against peace.

As Robin's two narratives about Monsanto demonstrate, pressure can be brought to bear on the reputation of mega-industrialists by means of a conjuncture of several elements: documentaries, international tribunals and a series of successful lawsuits. Indeed, it is surely no coincidence that Monsanto felt under pressure to initiate its sale to Bayer in 2016 (concluding the deal in 2018).[14] However, these unfavourable exposures are but a tiny part of the strategy necessary to change the model. Because, of course, Bayer's take-over merely serves to consolidate a larger share of the global agribusiness market for itself. Moreover, Bayer's own practices are not necessarily any better than Monsanto's – to that effect, it's worth looking at Bayer's recent webpage on Glyphosate, touting its 'marvellous virtues' with rhetoric linking it to biodiversity, soil preservation and safety for wildlife.[15] Nor indeed should we delude ourselves about the impact of recent major mergers into three companies of the former six giant players in agri-chemical-bio-technology: the already-mentioned German-American Bayer/Monsanto (2018); the American Dow/DuPont (2017); the Swiss-Chinese Syngenta/ChemChina (2017). These three companies own two-thirds of the world's seed market, have cornered the entire GMO-seed market, and control three-quarters of global agri-chemical sales (that is, Glyphosate and other herbicide, pesticide and fertiliser products). It is not necessary to explain how this massive monopoly puts the world's food supply in great danger. I have only to point to the GMOs – if all seeds are of the same type, diseases could quite easily annihilate them and a worse famine situation ensue.

Robin's two eco-docs expose how control of seeds by this handful of multinationals means that there is less choice and higher prices for famers. Indeed, these corporations have ensured their dominance by way of Intellectual Property Laws (IPL, patent and plant-variety protection laws) with the total endorsement of intergovernmental organisations such as the World Trade Organisation (WTO), the Free Trade Agreement (FTA), the EU, the UN and individual governments who have bowed to the pressures exerted. This control has allowed these corporations to establish a catalogue of proprietary seed sales which duly impacts upon existing indigenous seed species, thereby eliminating long-established crops, as well as fruit and vegetable products.[16] The case Robin presents here is neatly resumed in a nutshell by GRAIN (an international non-profit organisation supporting small farmers) as follows:

> Trade and investment agreements are a weapon of choice to impose seed laws where they did not exist before or to make existing laws more favourable to transnational corporations. The end goal is clear: to prevent farmers from saving seeds so that they buy corporate seeds on the market instead. And in that process, to get governments to pull out of plant breeding and seed production. … Trade agreements such as those required by the WTO and FTAs set market rules that supposedly aim to prohibit discrimination but may also give agribusiness preferred access to certain markets. As a result, governments may no longer be able to implement procurement programmes under which state

> authorities buy seeds from local farmers. (The rationale is that, by restraining competition, local procurement requirements put transnational companies at a trade disadvantage.) These are harsh conditions that give preference to corporations rather than to the welfare of farmers or consumers. ... It should be noted that United Nations agencies such as the UN Food and Agriculture Organisation, the UN Conference on Trade and Development or the World Intellectual Property Organisation are today important proponents of all of the above laws. They draft model laws and train governments in how to implement them.[17]

When, in 1969, by way of legitimising the use of Agent Orange – a Monsanto and Dow Chemicals product, by the way – during the Vietnam War to destroy food crops and deforest land, Henry Kissinger declared that 'food is a weapon', who could have imagined that half a century later we appear to be in a very similar relationship with food?[18] Not only have the former chemical agents, used in twentieth-century warfare, been re-morphed into pesticides, herbicides and fertilisers, the food chain is so affected by these dangerous elements of the much-heralded Green Revolution that all aspects of the environment including humans and other-than-humans are at risk, to say nothing of the many species of flora and fauna that have already been destroyed.[19] Killing as an agricultural growth-economy is really a bizarre outcome of humanity's ingenuity, until you stop to consider the huge economy of the war industry with which the agro-chemical industry is so closely identified – then, presumably, it seems less perplexing.

Embedded within these issues above, and linked to the agro-chemical industry's orders of economy, are GMOs. World-wide, 10% of arable land grows GMOs. In terms of hectare usage, the USA is the largest (75 million ha), followed by Brazil (51 million ha), Argentina (24 million ha), Canada (12.7 million ha) and India (11.6 million ha).[20] Just 38 countries ban GMO cultivation, 24 of which are European nations. These same 24 European countries, however, allow imports. A mere nine countries have imposed a complete ban (Algeria, Bhutan, Kenya, Kyrgyzstan, Madagascar, Russia, Venezuela and Zimbabwe). Presently, 26 countries plant GMOs (primarily corn, maize, soya and cotton). Of these countries, 19 are developing countries – and it is here in particular that ethical issues arise. The story unravelled by Robin's two eco-docs goes as follows. Since the mid-1990s, Monsanto successfully developed its GMO-seed range in tandem with its highly successful Roundup – these seeds were bio-engineered to be resistant to Roundup. Greatly assisted by the impact of the neoliberal globalisation project, which enabled the lifting of trade barriers, Monsanto was progressively able to market its double-package product (Roundup and GMOs) to developing countries who themselves had embraced the perceived benefits of economic liberalisation. First Monsanto offered their miracle Roundup as the solution to herbicide control. Then, to improve yields, Monsanto sold these countries on the merits of their GMOs (you can observe, on a mapping of Monsanto's areas of interest, a swathe of GMO crop areas going from Latin American countries, via

Southern Africa, up to India and across to other Asian countries[21]). But in signing up to this 'double-contract' of GMO seeds plus Roundup, these countries had to agree to buy new seeds every year – in other words, they were legally bound to *not* keep back any seeds from one harvest to the next (a process all traditional farmers pursue, as we know).

Robin investigates several specific cases of the Monsanto marketing strategies, beginning with what occurred in India with cotton farmers. Monsanto's first step was to convince the farmers of the merits of the GMO cotton-seed (as more resistant to disease than the native seed) and to sell cheaply to entice the farmers. As weeds became increasingly resistant to Roundup, so greater doses of spraying were advocated, representing a first increase in the farmers' budget. At the same time, more resilient GMO seeds were developed by Monsanto which meant these farmers, whose crops were failing, now had to purchase the more expensive Bt-cotton seeds, entailing a second increase in their budget. But, ultimately, these seeds were no more resistant to pests than the former ones, requiring more spraying, meaning a third budgetary increase. A last financial factor impacting on these budgetary increases was Bt-cotton crops' need for more water than traditional cotton crops, thus irrigation costs became greater. These high costs of cultivation (which often forced farmers to turn to loan sharks) and low returns left them caught in a debt trap from which there was no escape and, in turn, led many to committing suicide (around 17,000 per year). As the environmentalist activist Dr Vandana Shiva puts it, in *The World According to Monsanto*, these GMOs are no more than 'seeds of suicide'.

Both eco-docs show the consequences of this gradual process of increased expenditure for farmers who begin by trusting the marketing spiel Monsanto sells to them, only to find themselves locked into contracts they can neither extricate themselves from nor indeed afford to pay for. This creation of seed and monoculture monopolies, Shiva points out, destroys alternatives and is profoundly undemocratic. Indeed, a pattern emerges whereby Monsanto, through international trade agreements (which protect their interests basically), manages to assert, particularly on developing countries, the mass cultivation of its products. Thus, as witnesses from Mexico and Bangladesh explain in *Round-up Facing Its Judges*, they have found the introduction of GMOs (respectively, maize and aubergine) has either contaminated indigenous variants (maize), or actually failed as a crop (aubergine) when indeed the indigenous variants were not particularly susceptible to insects.[22] Shiva puts the Monsanto strategy perfectly when she declares: 'control the seeds and you control the world'.

A further two aspects of Monsanto's cynical campaign of seed-spread emerge in both films, and we are shown specific examples of this in *The World According to Monsanto*. One Monsanto strategy is using GMO cross-contamination as a way of getting into a country by the back door – as happened with maize cultivation in Mexico. Once Monsanto was able to prove the illegal presence of its patented GMO product on Mexican farmland, it was an easy next step to persuade farmers to accept the Monsanto industrial maize (rather than be prosecuted). Monsanto sweetened this enforced contract (at least initially) by selling it twice as cheaply as the indigenous variants of pure maize. But, by forcing the farmers' hand to turn to the GMO product and thus to spray with Roundup, their traditional corn economy and culture was put under severe threat.[23] The second strategy, as occurred with soya crops in Paraguay and other Latin American countries, was the unobtrusive introduction of GMO seeds which were shipped by Monsanto in white sacks (unlabelled therefore). Once the crops were established these countries had to accept a 'constat de fait' ('legal by precedent', as Robin puts it) and, moreover, pay royalties to Monsanto. Two types of Trojan Horse practice one could argue. Nevertheless, this seed-spread and pest-control industry is one endorsed as the Green Revolution by US governments and their agencies (beginning with the Reagan administration and through the two Bush eras to Clinton, as Robin's film reveals). American food-aid agencies and finance institutes quote Monsanto in their reports to the US House of Representatives Committee on Financial Services as a 'major player in agriculture' whose example should be followed since it brings the wherewithal to feed the poor. Yet, in the same breath almost (as if they can't help themselves from thinking in profit concepts), they also speak of the success of this Green Revolution brought about by Monsanto as something that 'has paid off handsomely'.[24]

The World According to Monsanto provides a devastating account of the collusion of the US Food and Drug Administration (FDA) with Monsanto over GM seeds.[25] According to the FDA no distinction can be drawn between a GMO plant and an ordinary one. Thus, the FDA's Principle of Substantial Equivalence applies and, as such, GMOs have been awarded Generally Recognised As Safe (GRAS) status. But such is not the case. To the FDA's statement that GMOs are a neutral technology (because they share the same DNA as ordinary plants) what they omit to consider is the gene exchange that also takes place at the laboratory level (genetic engineering[26]) – *these* are the additives that are *not safe*. But what transpires from Robin's interview with the then-Biotechnology Policy Coordinator of the FDA's Center for Food Safety, James Maryanski, is a clear signal of the agency's immutability. When confronted by Robin with the findings of scientific specialists that GMOs are not ordinary processes, and therefore not equivalent, and that the tests of the proposed GMO potato demonstrated that it induced digestive problems, Maryanski remained unconcerned (the GMO potato cells proliferated in the stomach of test rats).

It is worth mentioning, as does Robin's eco-doc, that Monsanto had actually funded the research centre, Rowett Institute, at Aberdeen University to carry out these potato tests (the potato was trialled and subsequently abandoned). But when the scientist, Dr Árpád Pusztai, went public with his findings on television he was immediately suspended by the institute, subsequently discredited. The funding was withdrawn and Pusztai's contract was not renewed.[27] It seems there is no stopping the massive engine of pressure that a multinational like Monsanto can exert, using all areas to push its products: its own status as a mega-industrialist, its lobbying of governments, its social networks which include the revolving-door practice of former Monsanto scientists, lawyers or CEOs being on governmental boards (FDA included) and vice-versa.[28] Equally, the corporation uses all means necessary to quieten any evidence that their products are harmful. That is the 'World According to Monsanto' which Robin exposes.

Even though GMOs are banned in Europe, Glyphosate (which includes Roundup) is not, and it is one of the most widely used herbicides (at 33% of the market share). It is used in more than 160 countries. As things stand, Latin America is the biggest consumer, followed by Asia-Pacific nations, then North America and Europe.[29] In Europe, France is the highest consumer and ranks as one of the highest users (along with Denmark, Poland, the Netherlands and Portugal) – this might explain why, despite promises to ban its use in France, President Macron keeps putting off the date. Its primary use is for annual cropping systems, perennial crops and grasslands. But it is its usage in the spraying of annual crops that is perhaps the most disturbing. First, it is used prior to sowing; second, shortly after sowing and finally, at the pre-harvest stage (known as pre-harvest desiccation) to quicken the ripening process of the crops and allow for a secure harvest in good weather conditions.[30] This triple-spraying practice is the one advocated for all annual cropping systems world-wide, and it is the effect of this high-impact spraying which has been linked to the carcinogenic dangers linked to Glyphosate.

Several health issues are raised in *Round-up Facing Its Judges* – which records and investigates the witness statements presented at the IMT. Rice farmers in Sri Lanka were not told how toxic Roundup was, nor how it would impact upon their health. Since the mid-1990s, when Roundup was first introduced, nearly half a million rice growers have developed chronic kidney disease which a toxicologist concludes is derived from exposure to Roundup spraying and from its leaching into the water supply (Sri Lanka banned its usage in 2014, and is the only country to have done so at the time of the IMT). Equally unaware was the Frenchwoman who sprayed her riding-school paddock with Roundup during her pregnancy, causing her son to develop oesophageal atresia in vitro with further complications of a tracheoesophageal fistula after he was born. A similar story of routine Roundup spraying causing this abnormal breathing condition is also brought to the tribunal by two mothers from an agricultural province of Buenos Aires. Roundup is a dioxin carrying herbicide (as is Agent Orange), and dioxin is a persistent organic pollutant and has been linked to several types of cancers, most significantly non-Hodgkin's lymphoma (but also birth defects, and digestive ailments).[31] Furthermore, of the two component parts of Roundup – isopropylamine salt and polyethoxylated tallow amine – it is the latter amine which, because it acts as a surfactant, sticks Roundup to the plant leaves being sprayed. This means that plants sprayed at the pre-harvest desiccation stage ineluctably contain high levels of Roundup at harvest time and that this chemical then, in a most immediate sense, enters the human as well as the other-than-human food chain. But, not only is there atmospheric pollution (the aerosol effect of spraying) that harms us and other-than-humans, there is also the fact that Roundup enters the soil (around 20% according to the expert witness Dr Dan Huber of Purdue University[32]). It subsequently leaches into water sources, and is harvested as part and parcel of a food crop.[33]

Yet, as the damning testimonies at the IMT and the documented evidence of *The World According to Monsanto* make clear, Monsanto has persevered with its rhetoric of success and used university research teams (usually funded by Monsanto) and scientists or governmental officials (often through some kind of revolving-door process) to publish untruths.[34] When you consider that 56% of Glyphosate-usage goes on spraying GMOs world-wide, 44% used primarily for standard crops, but also urban and domestic weed-control and that this market is predicted to expand by a further 6% in 2022, you can begin to see the enormity of the problem those who seek its ban are up against. Attempts to impose bans have been made across the five continents, but the outcomes are feeble to say the least.[35]

I will close this section with IDFA's chilling press release for French director, Emmanuelle Schick-Garcia's *Idiot Cycle* (2009), which reprises many of the findings exposed by Robin's two eco-docs, but which extends further the implications for our health by investigating the links between these agribusiness magnates and pharmaceutical companies. In her documentary Schick-Garcia exposes Monsanto, Bayer, BASF, Dow Chemical, DuPont and AstraZeneca[36] as the double agents they are, in that all produce pesticides, herbicides and GMO crops *and* produce and hold patents on drugs used for cancer treatment (a concerning circularity emerges):

> Consumers are being exposed to synthetic chemicals and genetically modified foodstuffs despite insufficient knowledge of the associated health risks. *The Idiot Cycle* links this exposure to the increasing number of cancer cases. In the Canada and United States of the 1970s, safety came first and products went through extensive preliminary testing. Nowadays, this step is often skipped. According to members of fact-finding committees, in a few cases negative results have even been kept secret or manipulated. Meanwhile, the producers of these chemicals and genetically modified foodstuffs (Dow Chemical, BASF, Monsanto, etc.) are also developing drugs for treating cancer, which provide them with huge profits. Are these facts related? And are consumers being used as guinea pigs? Director Emmanuelle Schick approaches the subject from a personal perspective – her mother has just been operated on for breast cancer. She highlights various aspects of the problem and invites numerous experts to discuss the issues. Where necessary, animations are used to clarify the sometimes complex subject matter. The picture that emerges is one of a conscious drive to make money at the expense of the environment and people's health. Furthermore, cancer research focuses more on developing new drugs than looking for potential causes.[37]

3.1.2 *Agribusiness in the dock – Part Two: Bee colony collapse disorder* **Vanishing of the Bees** *(2010) and* **More than Honey** *(2012)*

Not only does this lethal cocktail of Glyphosate and GMOs put humans at risk, there is also the impact it has on natural processes such as pollination. In the two eco-docs about bee colony collapse disorder (CCD), *More than Honey* (Marcus Imhoof) and *Vanishing of the Bees* (George Langworthy and Maryam Henein), we learn about other poisonous factors that enter into the landscape through spraying, causing harm to this most valuable of insects. These come in the form of the fungicide Ziram used to control leaf blight and scabbing on fruit, the insecticide Neonicotinoid (its most common brand name is Gaucho) and the herbicides Oryzalin and Paraquat. Ziram and Paraquat are the Swiss company Syngenta products (Paraquat is banned in Switzerland and the EU, and now the UK). Both products are highly toxic and, like Glyphosate, leach into the soil and water sources (causing other long-term problems). Paraquat (brand name Gramoxone) has been linked to fatal poisoning in developing countries (in Asia, Pacific Islands and Latin America) – how wise of Switzerland to ban its use in its own country of production therefore.[38] Ziram according to the FDA report is registered as 'moderate acute toxicity' with a 'suggestion of carcinogenic to humans' (some cocktail!).[39] Oryzalin is by DOW. It was deregistered, but then, in 2004, allowed by the FDA to reregister for industrial use only.[40] Neonicotinoid is a product mostly associated with Bayer. It is a neuro-active systemic insecticide; as such it affects the central nervous system of insects leading to paralysis and death. Its use has been restricted in Europe which has allowed bee colonies to revive. However, even where its use has been banned, that prohibition is always fragile. France banned its use in 2018, but that decision was recently reversed and the insecticide allowed back into the system for treating sugar-beet (the UK looks set to follow suit). Agencies such as the FDA and EPA in the USA, governments under pressure from agri-chemical multinational and farming lobbies, are weak and ineffectual when it comes to taking action. Worse still, as *Vanishing of the Bees* reveals, the EPA does not conduct its own research but asks the chemical companies themselves to do so; and these companies are not required to release these findings to the general public. This indifference is similarly in evidence at the FDA. Is it any wonder then that, when questioned in *Vanishing of the Bees* about the prevalence of these pesticides, one farmer merely responded 'if we didn't use pesticides, the insects would take over the world'.

All of the above chemical products impact on the well-being of bees (and of course other insects). As the two eco-docs explain, they weaken the bees' immune and digestive systems, attack the brain, affect their nervous system and cause navigation difficulties. According to scientists investigating CCD, bees may not immediately succumb to the effects of these poisons. However, when they return to their hives, they contaminate the hive through the chemicals they bring back home. GMOs (prevalent in the USA, Canada and developing countries) have, since 2003, had systemic pesticides introduced into the seeds which have been found to be yet another factor in reducing immunity in bees. Not satisfied with this level of interference, we learn and see how queen bees are genetically modified or artificially inseminated to meet the industrial needs of reproduction. Other cost-cutting exercises include depriving the worker bees of their honey and feeding them on water-syrup. When you consider that one-third of our food depends on bees, this level of abuse is startling.[41]

By exposing these facts, *More than Honey* and *Vanishing of the Bees* question the relationship between human beings and bees and the increasing conflict between civilisation and nature. Where, once, traditional beekeeping prevailed and the human could claim 'beekeeping is one of the few professions where you leave the world in a better place' (as one beekeeper puts it), what we witness is the industrialisation of the process, all with a view to making profit. In the USA alone, bees account for the pollination of $15 billion worth of food per annum. As one Californian almond orchard farmer proudly asserts: 'we're capitalists, we want growth, total global domination'. There you have it, and we see it in action. In *More than Honey* we witness the Taylorisation of bees as pollinators and honey-producers – literally, we see them being transported thousands of miles in trucks to do their work in orchards and farms; we see the hives' honey-supers (or boxes) being dragged along a conveyor belt, through a scraper mechanism to remove the honey whilst bees are still on the frames and crushed to death if they don't manage to escape (when all they'll get, if they survive, is water-syrup). It is all about money. *Vanishing of the Bees* tells pretty much the same story. We hear from industrial beekeepers who are caught in the system of their own making and cannot see a way to extricate themselves from what they do. They trundle their hives all over the USA, across 27 states. Typically, an almond orchard will demand 4000 hives for pollination purposes. For the beekeepers this represents $60,000 of revenue. So even if they lose their hives to CCD, they just rebuild their hive stocks and start all over again. In *More than Honey* the already mentioned almond orchard farmer declares 'I know the hives didn't mean to die. I've just got to move on'. He then, ruefully adds, 'my grandfather would probably walk away from this property if he saw how we keep bees today. He'd think "My God, you've lost your soul"'. And yet this farmer has no intention of changing what he does.

So the story is a familiar one – One Idea, One Culture, One System, One Deal – in this case, the artificial industrialised world of beekeeping, of worker bees pollinating monocultures rather than participating in the organic process of diversified agriculture in which they would thrive. Monocultures are a fertile world for diseases and pests, thereby requiring increased dosage levels in spraying – so of course this generates yet more revenue for the agri-chemical industry. Both eco-docs make clear that this mass interference with the natural process cannot be disassociated from globalisation. I'll just cite two examples shown in these eco-docs: that of bee importation and cheap imports. In a desperate attempt to curb CCD, beekeeping industrialists have turned to importing bees from other countries. Thus, California, for example, imported Australian bees. These are known to be more aggressive, and the fear is that they may outcompete the indigenous honey bee. No clear results have been recorded so far. This kind of practice should raise eyebrows, however. Brazil, for instance, imported the African honey bee, known also as the killer bee. This hybrid has now spread into the USA southern states and, as its epithet suggests, readily goes on the offensive if it feels threatened. But, as these eco-docs explain, it is this practice of bringing something bigger and stronger to the problem that is troubling, when the obvious fix would be to change the current practice of spraying, to put an end to developing monocultures that necessitate increased dosage of spraying, thereby preserving the indigenous bee. Even if these imported bees end up interbreeding with the indigenous honey bee who can say what the outcome will be?

In a fiercely competitive global market, a central tenet to the growth model which drives most economies lies in the belief of keeping prices low for the consumer. Thus, the American industrialist beekeepers argue they cannot compete with the cheap imports of honey from China if they don't use artificial methods. In *Vanishing of the Bees*, as if to underline the impact of this threat to business, we see huge vats of this imported honey which, as one industrialist beekeeper points out, 'is a scam'. These products are a honey-blend containing lactose, fructose and beet sugar. Moreover, because China's bees are anaemic they are dosed with anti-biotics, so that too enters our food chain. We also learn in *More than Honey* that China's bees are in sharp decline in certain provinces and that humans are having to do the pollination – when we consider that, since 2010, when China became (alongside Monsanto) a major producer of Glyphosate (Wynca and Xyngfa Group) and generic pesticides (ChemChina) we should hardly be surprised by this outcome.[42]

Both eco-docs make clear that there is a way to restore the bees' well-being. A first step would be to recognise that the industrialisation of beekeeping is not the way to proceed. Further steps that need to be taken are to offer diversified agricultural environments for bees to flourish, to ensure that they are properly nourished with their own product and that, as with the traditional method, the beekeeper takes only 50% of the hives' honey. Bees are selfless workers and their generosity deserves respect. Other steps include recognising the true value of bees by responding to the fact that it is essentially a feminine culture, as indeed *Vanishing of the Bees* makes clear. Ninety-nine percent of the workers are female and they do all the work (cleaning and guarding the hive, feeding the bee-brood, building the honeycomb, collecting the nectar); the queen bee is of course also female. The male drone bee's sole function is to fertilise the queen after which he dies (we see a spectacular example of this fertilising in full flight in *More than Honey* with a swarm of drones pursuing and mounting the queen bee). To recognise this feminine culture would surely be a powerful counter to the very masculinist culture of industrialisation and mechanisation that has developed around beekeeping and the concomitant growth in pesticide usage that mass monocultural pollination requires.

These two areas we have concentrated on, agri-chemical crop science and industrialised beekeeping, are just two aspects of the same pattern: a drive for global domination by multinationals which also entails imposing a closed-cyclical system which enriches only them (as identified in discussing Monsanto, but also noted with other chemical corporations, see *Idiot Cycle*). The centuries-old colonising mentality of the human species, which lasted well into the twentieth century, is far from being put to rest. Even if its militarised form no longer prevails, nonetheless, it perdures in the products made to control the planet's productivity through adapted chemical warfare technology. Let's see what the implications of this are by tracing what we have learnt from these eco-docs onto Raworth's Doughnut (Figure 3.2).

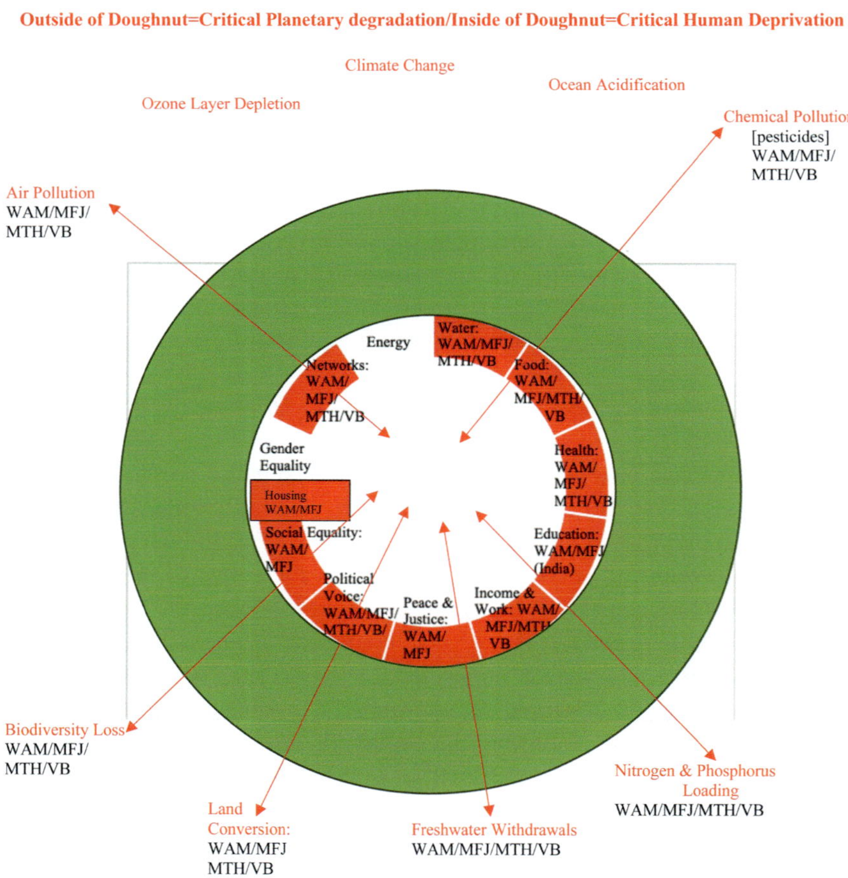

FIGURE 3.2 Mapping the agribusiness eco-docs onto the Doughnut (WAM = *The World According to Monsanto*; MFJ = *Monsanto Facing Its Judges*; MTH = *More Than Honey*;VB = *Vanishing of the Bees*).

To begin with the planetary boundaries: as you can see from the diagram, the eco-docs we considered make abundantly clear that at least six have been transgressed – and, undoubtedly, these will have a bearing upon the oceans' health, since rivers are affected by chemical run-off. As Kate Raworth explains, the impact of chemical and industrialised agricultural practices on the land reaches far and wide:

> Synthetic fertilisers containing nitrogen and phosphorus are being added to Earth's soils at more than twice their safe levels. Their toxic run-off has already led to the collapse of aquatic life in many lakes, rivers and oceans. ... Although the global scale of chemical pollution has not yet been quantified, it is of great concern to many scientists. And human pressure on other critical Earth-system processes – such as freshwater withdrawals and ocean acidification – continues to rise towards planetary-scale danger zones, creating local and regional ecological crises in the process.
>
> *(2017: 52)*

Raworth argues that until the economists' language shifts from 'calculating the monetary value of "natural capital" and "ecosystem services", ranging from the economic worth of the world's wetlands (said to be $3.4 billion per year) to the global value of insect pollination services (equivalent to $160 billion per year)' (7), biodiversity loss is going to continue on its severe trajectory (52). And the impact on humans, on the social foundation will be such that 'deprivations in life's essentials' will be worse than they already are. Currently, world-wide 'one person in nine does not have enough to eat; one in four lives on less than $3 a day, and one in eight young people cannot find work' (51). This is the outcome of the 'cuckoo goal of GDP growth' (60).

The diagram makes this perfectly clear. *Chemical pollution, phosphorus and nitrogen loading* combined with *land conversion* (to monocultures) have impoverished land, created *biodiversity loss, air pollution,* poisoned the planet's water – sometimes to levels dangerous to human health. *Food* and *water* are no longer reliably safe. The people most critically affected by the force of this industrialised agriculture are the poor – from the Latin American countries all across to Asia and most specifically the Indian subcontinent – all of whom are denied a *political voice* ('they won't listen to us'); excluded from *social equity* to the point (at its most extreme) they take their own lives, having lost their *homes* and their livelihoods. Children are obliged to work the land as free labour, so *education* is also in shortfall for them. *Networks* are abused to ensure the message of the chemico-agribusiness corporations is the one that carries weight with governments and agencies that are supposedly put in place to protect the environment and consumers. Even farmers who have bought into the mindset of these multinationals – such as the American soya farmers, the beekeeping industrialists or the French sugar-beet farmers – fall victim to the toxicity of this business model by insisting that there is no other way to go. Meanwhile the myth propagated by these companies that they are helping solve world famine is exposed for the untruth that it is. Most of the crops produced are destined for animal feed and those animals are bred to feed Europeans, North Americans and the Chinese.[43]

Two of the social boundaries not specifically pointed to in these eco-docs are *energy* and *gender equality*. In terms of energy and agribusiness, the focus, in the eco-docs discussed, has been more readily directed toward the *effect* of these products rather than energy consumption in the production and distribution of the chemical products per se (arguably, however, it is implicitly touched upon when we are presented with aspects of their global reach). On the question of gender equality and in terms of representation within these eco-docs, it is made clear that what remains exclusively male (and white) is the world of agribusiness corporations and their endorsers. However, in terms of those who are impacted upon, or who contest the value of the agri-chemical industry, all eco-docs present us with equitable gender representation, a multitude of voices and diverse ethnicities relevant to the countries represented.

To the argument that chemicals are the only solution to infestations and GMOs the end of famine, the case has been made that they are not; and that, therefore, the answers lie elsewhere, in a more holistic and diversified approach that puts the potential of what the Earth can carry at the heart of its process, namely, Agroecology. And it is to the eco-docs that examine these alternatives paths that we now turn our attention.

3.2 Regenerative system(s) of growth for the common-wealth

3.2.1 Agroecology at work: Solutions locales pour un désordre globale *(2010),* Voices of Transition *(2012),* Au nom de la terre *(2013) and* Demain *(2015)*

In an interview on French radio, the philosopher Bruno Latour asks the pertinent question: ‘can we overcome an industrial economy and prevent the ecological mutation?’[44] It is a sound enquiry when we consider how agricultural land is currently distributed. And I’d like to come back, first, to those statistics about the distribution of arable land I set out at the beginning of the chapter, and pursue them a little deeper, for they contribute to this part of the story. The 2019 UN *Food and Agricultural Organisation Report* (FAO) – calling for a ‘radical societal transformation upheld by food sovereignty’ – estimates there are around 608 million farms worldwide, 84% of which are small farms (up to 2 ha).[45] Most of these small farms are concentrated in Asia (China and India most notably). We recall that they represent a mere 12% of arable land however. So here is how the repartition of arable land to farm size pans out:

511 million small farms up to 2 ha = 84% of all farms, with a total share of 12% of arable land

61 million farms 2–5 ha = 10% of all farms, with 6% of arable land
30 million farms 5–50 ha = 5% of all farms, with 12% of arable land
6 million farms 50 ha up to 500 ha+ and above = 1% of all farms, but with 70% of arable land[46]

In Western Europe, farms over 100 ha dominate; in Brazil and the USA, farms over 500 ha. In Australia, the average size farm is over 4000 ha (the Australian farming business alone consumes 9.4 billion litres of water per year[47]). Large farm owners and the agencies that support them (such as the WTO, the EU Common Agricultural Policy, the FDA) will claim that large farms are more efficient. But, as the previous sections have shown, this is not the case in planetary terms. Furthermore, once the world is without oil how will these farmers manage to fuel their heavy machinery one wonders? Biofuel is not the answer. And what of water resources? The eco-doc *Lords of Water* (Jérôme Fritel, 2020), which came out a year after my specific corpus (2000–19), informs us that demand for water is massively on the increase, so much so that there is talk of a 'financialisation' of water, of what is known in the field of speculators as 'blue gold' (nice change from black gold wouldn't you say!). Banks, investment and hedge funds are already rushing to invest billions of dollars and euros into anything related to water. How remarkable we are: first, the gold rush; then the oil rush; now the water rush! But perhaps this last 'rush' is the most cynical of all. We know already that 2 billion people world-wide are suffering the consequences of water shortages. Estimates show that by 2050 this figure will treble.[48] As the experts in the film make clear, the monopoly of water has begun, but at what cost to the world's inhabitants and the Earth?[49]

It is clear that a deep economic, social and political transformation on a global level has to occur. It is clear also that the cultivation of mass monocultures is not helping famine, but rather, contributing to it. As the eco-docs *Solutions locales pour un désordre globale* (henceforth *Solutions locales*), *Voices of Transition* (the international release title for *Cultures en transition*), *Au nom de la terre* and *Demain* repeat in unison: 40% of our daily calorific intake *need* is going into animal feed and biofuel. Currently, around 820 million people go hungry.[50] A complete shifting away from the crops used for feed and biofuel towards direct human food consumption would feed an additional 4 billion people. Clearly, this 100% shift is an unlikely proposition. However, reducing the consumption of grain-feed animal products by 50% would mean increasing calorie availability to feed an additional 2 billion people.[51] The time has come in global history in which profit rights can no longer supersede human rights, nor equally importantly Earth's rights. And we know there is an answer. The planet can provide for us if we are willing to change our present agricultural productivity and food consumption practices (basically, producing and eating less meat). The evidence is there of the need to embrace change, to move towards reclaiming and regenerating land, making it healthy once more by diverse means such as permaculture, biodynamics and other organic systems of cultivation, reducing farm sizes (to an average mean of

250 ha) and practicing crop rotation, harmonising cultivation so that plants and other natural processes do the work of insecticides. Finally, a move away from the reliance on heavy farm machinery to lighter models would greatly assist in preventing the soil from compacting, as too would practicing conservation agriculture with reduced or no tillage of the soil and sowing cover crops in off-season (thereby reducing carbon monoxide and methane release from the soil). These combined practices would also ensure that employment can increase.

How to change our method is the focus of all ten Agroecology eco-docs. Of these, the four titles mentioned above will constitute the corpus of this last section of the chapter. All four have in common the message that it is time to build a new model by transitioning away from the current agricultural practices, recalling once again R. Buckminster Fuller's poignant words: 'you never change things by fighting the existing reality. To change something, build a new model that makes the existing model obsolete'.[52]

The names of the production companies involved in financing these eco-docs are also indicative of this underlying ethos of change. Colibris, the agroecologist Pierre Rabhi's foundation, co-produced *Solutions locales*, *Au nom de la terre* and *Demain*. Milpa Films, director Nils Aguilar's own company, co-produced *Voices of transition*. Both names are Amerindian in origin.[53] Colibris refers to the legend of the hummingbird (*colibri*) who – while others did nothing as a fire raged through their land – travelled back and forth to a pond with beakfuls of water trying to extinguish the fire. When ridiculed for this effort, the bird responded, 'I know, but I am doing my bit'. Milpa refers to an agricultural system used throughout Mesoamerica which is highly diverse. The method, commonly referred to as the Three Sisters, is based on the ancient Mayan tradition of planting three complementary crops together (maize, beans and squash) to provide for local subsistence. Taken together Colibris and Milpa indicate the path(s) to cultivate: rather than trying to control nature by subjecting it to scientific logic whose outcomes are driven towards profit, a far better method is doing one's part and doing one's best to protect the land (and thereby humans and other-than-humans) alongside using tried and tested practices that are in harmony with the land. Shared knowledge rather than shareholders!

There is, as you would expect, overlap in the subject matter covered by the four eco-docs when it comes to the alternative methods of farming discussed – all of which are organic and include permaculture, agroforestry, biomass and biodynamic farming. Similarly they tend to be structured around three dominant themes (these are French eco-docs after all!); however, they remain distinct in their approaches, and each one brings at least one new element into the discussion of transitioning to a new model (at the core of which beats the heart of Democracy). For ease of discussion, I've drawn a schematic outline to show how the eco-docs operate in terms of agroecology strategies they share, new elements they bring and the over-embracing concept of Democracy these strategies advocate (see Figure 3.3) which is deliberately mandala-like to make the point about the regenerative value of agroecology.

In the centre are the strategies these four eco-docs have in common (seen in the four quarters of the circle and the circular symbol). In the four boxes outside

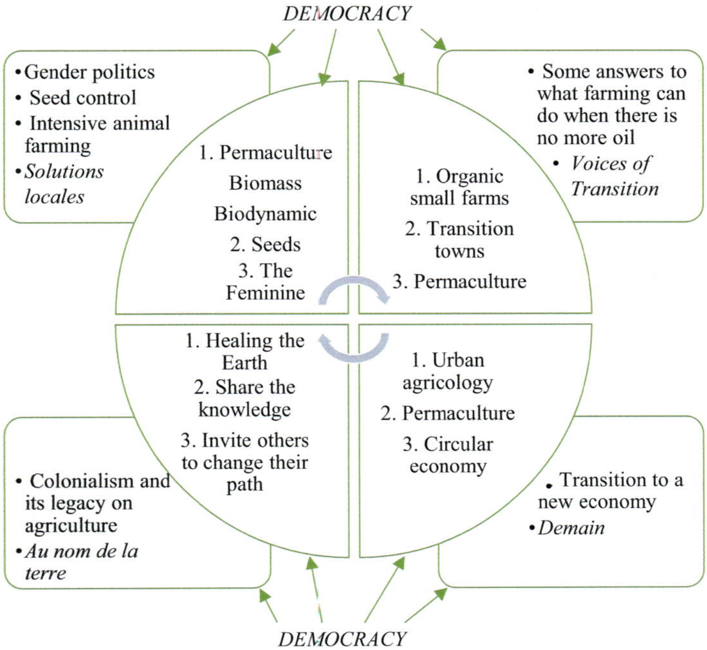

FIGURE 3.3 The regenerative and democratic value of agroecology.

the circle are the specificities each eco-doc brings. On the periphery of the diagram stands Democracy as the overarching ethos.

Before starting our discussion, it is important to acknowledge, first, some of the founder-members of these alternative methods, beginning with Rudolf Steiner's concepts of organic farming developed in the early 1900s. His biodynamic system advocated the inter-relatedness of soil fertility, plant growth and livestock care. This holistic approach is embedded in Steiner's spiritual beliefs which combine science and spirituality, and which he called anthroposophy. Permaculture, a term coined by Bill Mollison in 1978 (a conflation of permanent agriculture), is also a holistic approach to agriculture. It is a philosophy based on the principle of working with nature, learning from nature's ways (so biomimicry is part of its ethos) and includes aspects of Steiner's system such as regenerative agriculture, rewilding and community resilience (self-sufficiency). Permaculture works on the idea of a food forest – that is to say a functional and sustainable ecosystem with seven layers at different heights (trees, understory, ground cover, soil, fungi, insects and animals) – which allows a diversity of living organisms to grow in a relatively small space.[54]

Let's now return to our four eco-docs. As we shall see, although there is overlap, each one has its own tonality. Thus, *Solutions locales* feels much like a fencing match between the universality of agroecological solutions widely available and the current global agribusiness disorder. But it is an uneven match. For, as Serreau wryly remarks in her film, even though the UN's FAO stated, in 2007, that organic farming could feed the planet, not a lot has happened to allow this to be implemented. This is because, driven as they are either by their obsession with GDP or a belief in the value of mass-monocultures, Brussels, the WTO, governments, food agencies and, yes, even the UN (until its 2019 declaration cited above at the beginning of this section) all endorse (or lack the power to challenge) the industrial model and thus permit these 'pilleurs de la nature/looters of nature', as she calls the Agribusiness complex, to continue to commodify nature into an industry solely for profit.

Undaunted, Serreau, over a three-year period (2007–10), slowly documented the evidence which confirms the implement-ability of the UN's 2007 declaration. Serreau's investigation is spread across five countries in four continents (Europe, Latin America, Asia and Africa). Through interviews with specialists in agroecology and a broad spectrum of practitioners (agroecologists all, be they peasant-farmers in India, Brazil, Morocco, Burkina Faso or organic farmers in France and the Ukraine), her eco-doc demonstrates the efficacity of agroecology as a global response to the multinational corporations' globalised agribusiness practices. The first response is the 'healing of the soil', as Pierre Rabhi puts it. But, as he explains, a shift to the practice of agroecology – organic farming on a small scale – requires a change of 'mentality, and that takes time'. The best way, he believes, is to show by example. Rabhi began his agroecology work 60 years ago, developing methods on his own land in the Ardèche, beginning with making a healthy compost – what he terms 'humus' – to feed the earth. As he explains, the essential ingredients of a healthy small farm reside in biomass generation, recycling organic waste and rainwater conservation, plus an interactive combination of trees and shrubs with crops. Rabhi then took his methods to Morocco and Burkina Faso (amongst other countries) to show small farmers how to save their degraded soil and make their land productive once more, so as to become self-sustaining and beyond. We see him demonstrating how to make the compost, then dig it into the earth and how, once the soil is replenished, the use of straw serves as both a weed controller around the crops and a water retention device. The microbiologists, Claude and Lydia Bourguignon – who as early as the 1970s alerted authorities to the soil degradation (due to a loss of biomass and micro-organisms such as bacteria and fungi) – also developed earth-healing practices based in organic farming and the integrated biodynamic crop-livestock system. Much along Rabhi's lines, they advocate (and practice) an agro-forest-pastural type of agriculture. As Claude Bourguignon – holding, in one hand, the compacted soil ('like cement') of industrial farming practices and the friable organic soil ('like couscous') in the other – neatly puts it: 'the land, the fields, trees, animals, that's proper agriculture'.[55]

The second thrust of Serreau's film concerns the various resistances to industrial methods. One such case is the Indian farmer who managed to get out of the 'profit game', abandoning the industrial method and reverting his 1.8-ha farm back to traditional organic practices. This enabled him to become not only self-sustaining and able to produce enough to feed his family and have some to sell on, but also to have the wherewithal to employ local people to help out on the farm. Also in India, the director of the, then, Kokopelli-funded project (now Annadana) leads us through the practice of biomass (in this case, growing grass high, then cutting it down and covering it with soil), biodynamics (making fertiliser from cow products and fishmeal extract from mackerel) and biopesticides (using ginger and turmeric) as employed on a rice farm – again showing the way for small farmers (the dominant category in India) to succeed and not be trapped in the industrial model. In Brazil, where concentration of land ownership means that two-thirds of arable land is in the hands of 3% of the population, the landless peasant-farmers movement in Brazil, the MST (Movimento das Trabalhadores Sem Terra/Landless Workers Movement) has stood firm in its claim that it is legally justified in occupying unproductive land that does not meet its social function (as per the Land Reform Act of the 1988 Constitution). Some 1.5 million rural poor belong to this movement with just 31,000 families occupying land. However, despite the Constitutional declaration, as the couple interviewed by Serreau confirm, they are never far from the threat of violence and eviction from the landowners. It is a perilous life they lead, yet one whose only goal is to refurbish the land and be self-sustaining. We also learn that this movement is not just opposed to the landowner, but also to agribusiness – or capitalist farming as the couple puts it – which they see as the organisation of industrialised agriculture brought in by foreign capital and products (in particular GMOs, along the lines we saw in Section 3.1 of this chapter in relation to Monsanto).

Finally, amongst these resistances portrayed is the fight against the seed domination exercised by the three consolidated multinationals, Bayer/Monsanto, Corteva (DuPont-Dow) and ChemChina (Syngenta-ChemChina).[56] Ever since the 1930 Plant Patent Act (USA), enacted in response to the growth in hybrid seed production and to protect the producers' intellectual property rights, agribusiness has progressively extended its tentacles to the point that, as both Dominique Guillet, the founder of the 'Seeds Without Frontiers' Association Kokopelli, and the organic farmer-researcher Philippe Desbrosses point out, an Official Seed Catalogue has been established requiring seeds to be certified in order to have licence to sell – this also includes the prevention of seed retention for the following year's crops (because, as these patents decree, contamination could occur). The bodies governing these licencing laws are of course in the purview of the three giants of the seed industry named above.[57] By 1995 – effectively since GMOs and Roundup took a stranglehold – and once the WTO was established, this virtual monopolistic practice was set in stone, as GRAIN explain:

> Ever since the establishment of the World Trade Organisation, and almost without exception, all countries of the world have passed laws giving corporations ownership over life forms. Whether through patents or so-called plant breeders' rights or plant variety protection laws, it is now possible to privatise micro-organisms, genes, cells, plants, seeds and animals.[58]

As food-sovereignty advocate Dr Vandana Shiva tells Serreau, trade and investment agreements and the Agreement on Trade-Related Aspects of Intellectual Property Rights (TRIPS) constitute a form of 'biopiracy' whereby farmers are prevented from saving seeds so that they are obliged to buy corporate seeds instead, with the knock-on effect that governments pull out of indigenous plant and seed production. The impact on poor and rural peasant farmers is disastrous. The problem, also, is that these patented hybrid seeds, by dominating the market, reduce the number of varieties available. Claude Bourguignon relates how some 3500 apple-types in France are not recognised by the Catalogue and thus are rapidly disappearing, replaced, for example, by the tasteless Golden Delicious (which requires lots of pesticides).

But, as Desbrosses explains, a legal loophole was found to counter this monopoly. By setting up farmer cooperatives and other types of associations, non-industrial seeds could be licenced from one farmer to another. Since the late 1990s, a number of cooperatives and seed banks world-wide have been established, three examples of which we visit in Serreau's eco-doc. We witness a cooperative in Brazil of 12 farmers producing organic seeds and sharing them out. We visit Dr Shiva's Navdanya Organisation in India (established 1998) which now has 150 community seed banks over 22 states. Seeds are given to farmers – one kilogram. Upon harvesting, they either return two kilograms of seeds to the bank or pass the two kilograms onto two other farmers (and so on).[59] Finally, we go to Kokopelli's headquarters in France. Established in 1999, Kokopelli licences seeds free of rights and reproducible to farmers all over the world. The guiding ethos is to support agroecology, protect biodiversity and assist poor farmers to gain seed autonomy.[60]

As Shiva points out 'seed is a gift' and should not be treated as a commodity. This idea of ownership, she explains, which guides the thinking of the likes of Monsanto ('own the seed and you have full control'), is just another facet of the masculinist approach to agriculture. Agribusiness seeks to dominate and control Mother Nature (to the extent of nearly killing her). Yet agriculture, as the Austro-Brazilian agronomist Ana Primavesi asserts – echoing Shiva's sentiments – is feminine. The Earth is feminine. For women, seeds are not merchandise but nutrition for all. The commodification of food, Shiva remarks, deprives us 'of our commonwealth'. It also affects *gender equality* – for, women, too, are seen as commodities. In India, 80% of farm work is undertaken by women, yet they only own 13% of the land. Worse still, Shiva adds, females are so devalued in India, which has the highest incidence of female infanticides – baby girls are seen as a burden.

The three other eco-docs have much in common with the premises set out in Serreau's film. So, rather than re-iterate solutions already proposed, I'll now focus on what else they bring to the agroecology table. Much of Pierre Rabhi's philosophy and practice is already explored in Serreau's film – which is hardly surprising, since, in 2003, she had embarked on making a documentary about him which then grew into the much larger project of the film she finally produced,

Solutions locales. However, in the eco-doc *Au nom de la terre* (Marie-Dominique Dhelsing), we do obtain a fuller picture of his anthroposophical approach and a deeper appreciation of this man on a mission to heal the Earth. Dhelsing's film traces Rabhi's 60 years of commitment to a 'land of humanism', to a social lifestyle of 'happy sobriety' rather than 'over-consumption'. Rabhi explains how, in 1961, he established his own farm, aptly named Montchamp (my field, or field on a hill); he shows us his working methods (organic humus production, permaculture, biodynamic agriculture, biomass) all of which contribute to the slow healing of the land. We then follow him to Gorom-Gorom in Burkina Faso. Situated in the northern province of Oudalan, this town is an important cross-road in the Sahel – which is itself now almost a desert. Rabhi introduces his methods to the local farmers to help them save this 'degraded earth'. He passes on his knowledge of land management that includes agroforestry and permaculture – in short, he delivers a 'rewilding of the soil'. He explains about the symbiosis of plants, the fact that plants bargain with each other and so sustain each other, adding that we need to learn from nature and the value of biomimicry. After all, as he says, 'seeds are the propagation of life, without them humanity disappears'. By reclaiming the land for organic agriculture, Rabhi shows how self-sufficiency can be obtained and preserved by re-investing profit into the land and its people. By showing how agroecology works and that change is possible even with very little means, Rabhi invites the farmers onto the path of self-sustainability.

The historical-political impact value of this part of the eco-doc in Burkina Faso should not be lost upon us. Rabhi explains how, under President Thomas Sankara (1984–7), this former French colony experienced an ambitious socio-economic programme to redress the effects of both the colonial (1890–1958) and post-colonial eras (1960 to the early 1980s). Amongst these reforms, land was redistributed to peasant-farmers and organised along the principles of self-sufficiency. The tentacles of colonialism run deep, however. A committed pan-Africanist, Sankara's policies were an aggravation to the then-French government who preferred Blaise Compoaré as an amenable presidential candidate.[61] Thus, this short-lived successful programme ended with Sankara's assassination, when once again, under Compoaré, the country reverted, as Rabhi puts it, to a politics of 'greasing palms that led Burkina Faso to give itself over to land-exploitation of resources and GMOs' (in the former case primarily gold, and the latter, Bt cotton). This, Rabhi tells us, led to a 'huge dependency on super-corporations who could care less about people' which, in his view, amounts to no less than 'a crime against humanity'. As a result, what currently prevails is subsistence farming and high levels of food insecurity.[62] With 40% of the population registered under the poverty level, unsurprisingly, migratory flows are increasingly a matter of everyday life. Burkina Faso remains a country that is rich in gold (the fourth largest producer in Africa), but unable to feed its people.

The tone of Dhelsing's eco-doc – perhaps because it follows the trajectory of an individual pioneer in agroecology – is expository in that it shows the why, the how and the value of Rabhi's practice. The drive and energy of this man to fulfil his prayer, which translates from the film's title as 'In the Name of the Earth' or 'For the Sake of the Earth', does not mark him out as a zealot but as a deep believer in the purpose of his project. His Earth-humanism has reached many parts of the globe. Closer to home: his daughter continues his ideas; a business-man, in the Drôme, rethinks his life and, recognising the importance of the younger generations' future, sets up a Centre for the Transmission of Knowledge; an aristocrat gives over her château and land for artists to create and for agroecologists to rewild.

Such commitment does not come without hardship, however, as Rabhi's wife, a determined supporter of his approach, lets us know. She says, 'it was tough going, but we followed a dream' – a sentiment that the eco-doc *Anaïs s'en va-t-en guerre* echoes, incidentally, in which we see this young woman's fierce determination to fulfil her dream to become a top restaurateur's choice for her herb cultivations (despite all the local administration's scepticism). But the point is also, as the closing moments of Dhelsing's eco-doc make clear, that we all, irrespective of our means, can contribute to this healing.

Nils Aguilar's *Voices in Transition* (2012) speaks of cultures of change (referring of course to the original French title *Cultures en transition*) and provides us with three stories about food sovereignty all of which are based on the practice of heterogeneous agriculture. At their core lies the concept of a circular, regenerative economy made possible by moving away from industrial agricultural practices to small-scale decentralised production methods. First, Aguilar documents the practices of small organic farmers (agroforestry, apiculture, permaculture); then, community-led solutions such as transition towns and urban agriculture; finally, Cuba's path to agroecology when faced with the loss of cheap oil as a result of the dissolution of the Soviet Union (in 1991).

The first story takes place in France. In answer to the EU's CAP policy, whereby 85% of agricultural aid goes to 17% of the biggest farms – thereby encouraging the concentration of labour and monocultures which increases both pollution and local rural unemployment and, furthermore, means that prices are fixed outside the farmers' control – there is an evident need, says one young organic-permaculture farmer (Oswaldo Forey), to move to a transition model of interdependent reliance. This means, he explains, 'enfranchising ourselves from our slavery to industrialised agriculture, practicing agriculture on a human scale, with each one of us bringing our expertise and sharing our knowledge'.[63] Crucially, money must be kept within the farming communities, he adds, thereby creating a circular economy that guarantees freedom from the multinationals.

The second story brings us to the UK and to examples of transition towns and urban agriculture on a small scale. Bob Hoskins, the founder of transition towns which was initiated in Totnes, Devon, explains how we can move away from excessive consumerism and still keep a town or city vibrant, as Totnes has done, by introducing its own currency. Furthermore, by investing locally we can fund local systems. Transition towns are committed to becoming more energy efficient by using mills and water courses (rivers, etc.) as sources of energy, by making transport (of food in particular) less carbon-footprint heavy and, wherever possible, by facilitating the ability to work where we live. Hoskins recognises these endeavours require establishing a plan and bringing in expertise to help implement it. As of 2012, when this film was made, there were, according to Hoskins, 1100 transition initiatives in 43 countries preparing their economies for the shocks to come. Bristol is one of those transition initiatives, and in Aguilar's eco-doc we visit a number of examples of urban agriculture, from a largish concern, the peri-urban Leigh Court Farm which practices permaculture to grow food within a community scheme (selling local, sharing knowledge as well as teaching skills), all the way across to small front-gardens growing produce to share with neighbours.[64]

Urban agriculture is, then, a further example of the practice of interdependent reliance, this time by teaching people how to grow produce on a community basis. The third story expands on this practice by taking us to Cuba to witness the resilience (also a transition term) people showed when faced with the impact of being a country without oil. The collapse of the Soviet Union pushed Cuba into having to become self-sufficient in feeding itself. Under Castro, Cuban agriculture flourished because of its dependence on Russian oil. But it was an agriculture that was industrialised and composed of large farm state cooperatives with concentrated monocultures. When the oil stopped, Cubans were faced with the danger of famine and so turned to urban agriculture. This was a gradual process, beginning with Havana's scheme of organic urban gardening. Everything is hand-managed, the horticultural-agronomist Samura Torres explains. Produce is healthy. Planting is balanced, including the use of plants to protect crops (based on the Three Sisters principle). Twenty years later, at the time of this eco-doc, 500,000 Cubans were participating in organic urban gardening and providing for 70–80% of what they need to eat. Progressively, the large cooperatives have been broken down into self-managed cooperatives (in 2012, only 40% of the land remained to be regenerated). Animals are used to plough the fields, goats and rabbits for cheese and meat. This decentralising of agriculture has led, we are told, to 150 cooperatives now feeding their own districts with organic produce and receiving double the typical salary.

Good food and economic autonomy, that is the ethos of Transition Culture 'which means a better way of life, more time, a way to celebrate life', concludes Bob Hoskins at the end of Aguilar's eco-doc. To achieve this on a global scale two crucial things need to be achieved, however. First, as the bio-economist and agronomist Professor Christian Dupraz explains, is to get agriculture out of the WTO.[65] As he says, 'the WTO's vocation is contrary to food sovereignty. The WTO is targeting a totally open market and the elimination of import protections'. Second, as Dr Vandana Shiva asserts at the very beginning of the eco-doc, local economies must free themselves from 'their forced integration into an international economy controlled by speculative monopolies'.

Three years after Aguilar's contribution to the transition debate and the need for new systems of economies, Cyril Dion and Mélanie Laurent released their extremely successful eco-doc *Demain* (2015), which – at almost twice the length of Aguilar's film and more in the tone of a 'road movie' (as Laurent puts it) than tightly focused – reprises many of the elements of *Voices of transition*. There are no audience figures for Aguilar's film which had a good showing at festivals (over a five-year period, 2012–17). It was also screened free at many venues world-wide and inspired a number of transition initiatives.[66] For its part, *Demain* garnered an $11 million box office and won the César for best documentary in 2016. The three thematic elements of *Demain* (urban/peri-urban and rural agroecology, permaculture and circular economies) take into account a number of transitional issues, involving ten countries, as they affect agriculture, energy, economy, democracy and education. Food sovereignty is at the heart of the agroecology and permaculture stories told here which investigate how people can produce their own food without recourse to transport, how the self-management style of employee-ownership organic farms functions (including coordinating deliveries, thereby reducing dependency on oil). Educating and training future generations in holistic ecosystems of planting and cultivation is explored, showing how even micro-farms (1 ha) can be efficient and productive of up to 60% of our dietary needs.

Food sovereignty and economic autonomy go hand in hand in this expository eco-doc which invokes the mantra of use-recycle-regenerate as the path not only to good economic practice *per se* but also toward zero waste (be it in the domain of energy or food production). And of course this economic principle is completely the opposing model to that fostered by the neoliberal model that hunts everlastingly after more growth. To that rather unmeaningful sentence 'let's grow the economy' (of GDP-land) stands this egalitarian idea of local economy models which can then be networked globally. Dion and Laurent's eco-doc supplies a very clear explanation of these two distinct models as follows. The problem of money is how is it created and why? It is created by banks for the purpose of profit via loans. When that money is paid back, with interest, that original money/loan no longer exists, only the interest paid. So, if we don't borrow, then there is no money. Natural ecosystems never allow for a monoculture – such as money, for example. So, as with ecosystems, which require varied species to survive, economic biodiversity is also required within the economic sphere. Economics (derived from the Greek *oikos nomos*), as Raworth says, means 'household management' (2017: 4). A household, the land, the planet Earth, all necessitate a pluricultural approach – we live in a world that is eco-systemically interdependent in which all is connected. Nature teaches us this. Thus, as *Demain* points out, some of the ways in which we can shift away from the current GDP-growth model is to think in terms of circular economics. As Bob Hoskins explains, one such method is the concept of keeping the economy circulating within the town or city. A number of places are already are doing this. Thus, instead of contracting outsiders to come in and do the work, city and town councils are employing and buying materials locally. Another method, on similar lines, is the introduction of 'complementary money' or transition money (such as the Totnes or Bristol pound) – a currency that has no value outside of the town or city, but which can be used within the community. Banks can't touch it because it has no value; no interest can accrue on it, so there is no drive to accumulate wealth. Instead, it allows communities to survive outside the central bank economy. This economic system is based in trust, a collective investment in community and the exchange of goods and services – a renewal of the social contract as it were. Moreover, it stops larger companies coming in and allows small, local companies to grow and produce; it also lowers costs of delivery since production takes place within the area.

The *solutions* provided by these four eco-docs mean that in terms of the Doughnut economic model all 12 of the social dimensions are met and all 9 planetary boundaries respected. In terms of actual *practice*, however, several ongoing issues still require resolutions, most significantly the case of *gender inequality* and therefore *social equality* in the farming context of India. Abuse of power by landowners in Brazil is also shown to be a hindrance to self-sustaining small-farm agroecology. Thus, the lack of a *political voice* and *peace and justice* still remains a critical issue for those peasant farmers who are mistreated, in contravention of Brazil's Constitutional Law. What remains clear throughout however is that the knowledge is there to change the existing systems, the will is there to share this knowledge, and there is evidence from all parts of the world that these agroecological solutions work and constitute the path towards healing the Earth and keeping humanity fed.

As in Chapter 1, we can observe how this drive towards regenerative practices in all ecosystems constitutes a direct challenge to global capitalism and its singularly undemocratic processes. Where is democracy, these eco-docs demand, if there is no power to change the laws and trade agreements that favour the wealthy (multinationals, shareholders and all)? Without agricultural autonomy there is no equality and, therefore, no democracy. If both the laws of Gaia and the Human Rights Act are to be respected, then the same triangular dynamic, already discussed in Chapter 1 and illustrated here in Figure 3.4, must be implemented.

Clearly, to enable this transition promoted by agroecology, it is crucial that economists start talking in the same vein. Thankfully, a number of economists are thinking in ecosystems and 'making complexity economics, network theory and evolutionary economics among the most dynamic fields of economic research' (Raworth, 137). They have immersed themselves in 'real-world economic

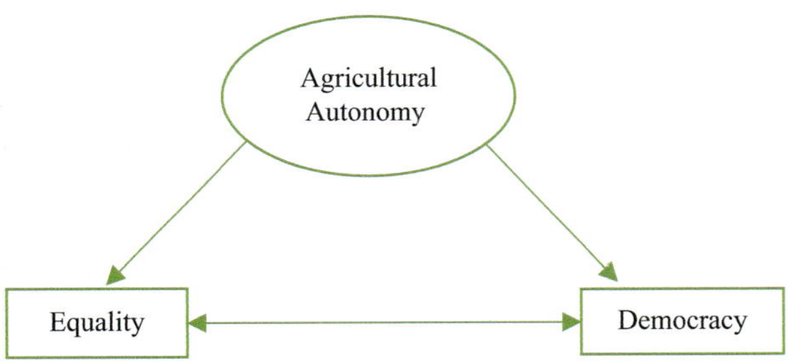

FIGURE 3.4 Agricultural autonomy as a source of equality and democracy – as promoted by agroecology.

challenges' and sought out economic thinking that will serve humanity and the planet (9). This is the commitment Kate Raworth's *Doughnut Economics* engages with. Raworth's Doughnut model, in its demonstration of the viability of a circular regenerative economy, connects with many of the elements about agroecology and its challenge to agribusiness as seen in the eco-docs discussed above. She speaks of the impact of the coming end of oil on food prices. It will lead to higher prices, she asserts, and to a 'drop in *industrial* output and food production, ultimately resulting in famine' (155, my emphasis). We are, of course, already witnessing this, as Dr Shiva says (each day 25,000 people die from hunger or malnutrition, 10,000 of them children[67]). Within this industrialised world of agribusiness, Raworth adds, it is not just a case of a seed monopoly (which we've already discussed), but the fact that grain is also in the hands of the mighty few – the ABCD commodity traders group (ADM, Bunge, Cargill and Louis Dreyfus) which controls over 75% of the market (148). Food is financialised and commodified to our detriment, a profoundly undemocratic state of affairs (Dr Shiva repeatedly remarks upon this in *Solutions locales* and *Voices of Transition*). Thus, we find ourselves in a system where 'global economic development is caught in the twin dynamics of growing social inequality and deepening ecological degradation' (154).[68]

Raworth also points to the malfunctioning of the money-monoculture, an issue Bob Hoskins discusses in these eco-docs, and she argues the case for different kinds of money to co-exist 'with the potential of turning a monetary monoculture into a financial ecosystem' (183). This would mean bringing our economic activity within the green biospheric space of the Doughnut where it would operate between the social foundation and ecological ceiling – the safe and just place for humanity. Several ways are to be found, she remarks, for this monetary redesign – most of which, she acknowledges, would need to be state-led. She proposes the four following ideas. First, 'separate the role of providing money from the role of providing credit' (183). To accomplish this, banks would

> take back the power to create money and then issue it to commercial banks, whilst simultaneously requiring them to hold 100% reserves for the loans that they make – meaning that every loan would be backed by someone else's savings, or the bank's own capital.
>
> *(183)*

There are already several examples of this practice in the form of ethical banks (Triodos for example) and (non-state) microfinance companies such as the Shared Interest Society, or the Grameen Bank (mentioned in Chapter 1).[69] Second, get 'State-owned banks to channel substantial low- or zero-interest loans into investments for long-term transformation, such as affordable carbon-neutral housing and public transport' (183–4). This latter system could also extend to agriculture and land restoration. Third, establish 'complementary currency for use by small businesses' within communities. Finally, tackle 'deep recessions by issuing new money directly to every household as windfall cash to be used specifically for paying down debts' (184). This distributive process could also work, says Raworth, by channelling a proportion of overseas development assistance (ODA) 'directly to people living in poverty in those countries' (199). It's worth suggesting, also, that if we moved away from the growth-economic model then it is highly probable that deep recessions would not occur. Finance would not be about speculation which is about buying and selling hype (an illusion, as Gordon Gekko in *Wall Street* puts it), but about investment in Earthly realities.

It is heartening to learn that, in such a short span of time since its launch in 2012, Raworth's Doughnut model has been embraced by around 500 communities worldwide (a number that is on the increase). Because her model is based upon a coherence between economic policy, environmental and social issues, it is malleable enough to be adapted to a specific city or country's own situation (for example Costa Rica and the Netherlands, two early examples of implementing Doughnut practices which obviously address different conditions, or more recently, Senegal's ecovillages).[70] Significantly, echoing Rabhi's Colibris Foundation's view of each community contributing to Earth's well-being according to their means, the Doughnut model indicates the ways we can take action on a local, national, international and global level to create a safe and just space for humans and other-than-humans alike – designing and implementing pathways to a sustainable future. In all instances, the point is to share the value of the commonwealth from the start by developing activities that are distributive and regenerative by design. As Raworth puts it: 'how many kinds of value can (an individual, a company, a community, a country) integrate into its design to make sure that (it) can give back to society and the environment?'[71]

As things stand, however, and in relation to agriculture, a predominant proportion of the rural poor are without land rights which would enable them to self-sustain. Many farmers struggle to eke out a decent living. Numerous countries in receipt of ODA continue to flounder and fail to feed their poor. Global warming continues on its upward curve (what an ironic truth when we consider the obsession of economics with the growth-curve model!) rendering soil uncultivatable – all of which only serves to increase the migratory flow condition. This latter sad state of affairs will be the central focus of the next (and final) chapter, which is on the effects of disrupted ecosystems on humans.

Notes

1 I'm somewhat rephrasing David Harvey's 'accumulation by dispossession' (2019). For more information on this economic concept see: <https://www.democracyatwork.info/acc_accumulation_by_dispossession> (accessed 24 October 2020). Here is what this amounts to in practice – I am quoting from the Permanent Peoples' Tribunal of Bangalore, 2011, which took to task agricultural multinationals and related their environmental harm to the 12 dimensions of the social foundation: 'During the last decades a corporate system has been set up, which is based on so-called "accumulation by dispossession" (David Harvey, 2019): appropriation of returns on capital by exploiting people and nature, violating human rights and disenabling people, by disregarding and impairing nature, thereby undermining the capabilities to create a humane future.

The repercussions on human rights are disastrous as nearly all witnesses explained. The effects are so serious because the economic activities of agrochemical Transnational Corporations *undermine all dimensions of human security*: environmental security, socio-economic security (as defined by the International Labour Organisation), health security, food security and safety, shelter, public security and also political security. Without these securities human development, as it has been conceptualised by United Nations Development Plan since the beginning of the 1990s, is not possible' <http://permanentpeoplestribunal.org/wp-content/uploads/2011/12/TPP_Bangalore3Dec2011.pdf>, 14 (accessed 20 October 2020). The tribunal, which predates by six years the case brought against Monsanto as recorded in Robin's documentary, concluded on the evidence provided (which is virtually the same set of arguments as presented to the IMT) that the agribusiness corporations Monsanto, Dow, DuPont, Bayer, BASF and Syngenta were 'responsible for gross, widespread and systematic violations of the right to health and life, economic, social and cultural rights, as well as of civil and political rights, and women and children's rights'. Moreover, the tribunal stated that it considered 'the home governments of the corporations complicit with their respective agrochemicals in all the violations'. Indictments as we know, but no action, as the 2016–17 Monsanto tribunal makes clear. For, Monsanto, though found responsible, rejected the whole proceedings as a 'mock trial' (see: <https://www.tandfonline.com/doi/full/10.1080/14747731.2019.1592067> accessed 23 April 2021).

2 This mean I have no space to discuss *Darwin's Nightmare*, an eco-doc that supplies a visualisation of the impact of colonialism, post-colonialism and Europe's current global practices upon the planetary and social ecology. As the title suggests, we are presented with the evidence of an evolutionary process that has gone horribly wrong. That process is centred around a fish, the Nile perch, which was introduced into Lake Victoria's waters in the 1950s, during the period when Tanzania was under British colonial. This powerful predatory fish ended up depleting the indigenous fish stock (a mainstay of the Tanzanian people). In so doing it also decreased the number of algae-eating fish, allowing the algae to grow significantly, to the point of choking the lake. Finally, it transformed the economic ecology of the local fishing industry. I have written about this film in Hayward (forthcoming 2022) , 'European-documentaries as Performative Exchange', in: Gábor Gergely & Susan Hayward (eds), *Companion to European Cinema*, London, Routledge.

3 See: <https://ourworldindata.org/land-use≥ (accessed 20 October 2020).

4 See: the most recent report from the UN issued in November 2019 'Food and Agricultural Organisation' <http://www.fao.org/3/ca7036en/ca7036en.pdf> (accessed 17 November 2020). See also: <https://www.fcrn.org.uk/research-library/how-much- food-do-smallholders-produce> (accessed 20 October 2020).

5 A commonly held view that small farms produce 70% of our food is in fact incorrect. See: <https://www.sciencedirect.com/science/article/pii/S2211912417301293> and <https://www.resilience.org/stories/2019-08-21/small-farms-dont-produce-most-the-of-the-worlds-food-but-they-could-produce-all/> (both accessed 17 December 2020).

6 See: <https://www.google.com/search?rls=en&source=univ&tbm=isch&q=pie+chart+for+worldwide+land+degradation&client=safari&sa=X&ved=2ahUKEwits-jnhMjsAhVStXEKHUB7AjQQjJkEegQIChAB&biw=1591&bih=941> and <https://www.ielts-exam.net/academic_writing_samples_task_1/985/> (both accessed 20 December 2020); with regard to the second reference, at least these issues are being taught in classrooms!

7 See: <https://www.twinkl.com/illustration/global-water-resources-geography-diagram-pie-chart-secondary> (accessed 20 October 2020).

8 See: <https://ourworldindata.org/land-use> (accessed 29 October 2020). Most recent figures suggest forests now only constitute 31%, see: <http://www.fao.org/state-of-forests/en/> (accessed 29 October 2020).

9 Source: <https://ourworldindata.org/global-land-for-agriculture> (accessed 23 April 2021).

10 See: <https://ourworldindata.org/world-population-growth#how-has-world-population-growth-changed-over-time> and <https://www.prb.org/2020-world-population-data-sheet/> (both accessed 20 October 2020).

11 This very much depends on what is counted as urban environment. Thus the UN puts the urban population at 55%, whereas the European Commission puts it at 85% (see: <https://ourworldindata.org/how-urban-is-the-world>, accessed 20 October 2020). Elsewhere claims are as high as 95%. What is clear is that urbanisation is on the increase world-wide.

12 See: <https://ourworldindata.org/urbanization> and <https://ourworldindata.org/global-land-for-agriculture> (both accessed 20 October 2020).

13 <https://www.telegraph.co.uk/news/earth/agriculture/food/12171193/Majority-of-supermarket-meat-comes-from-animals-fed-on-GM-crops.html> (accessed 27 October 2020).

14 Where Monsanto is concerned these include the use of Agent Orange, PCBs leaching into soil and water, chemical poisoning from Roundup spraying, the spread of experimental glyphosate-resistant wheat, the herbicide Lasso, the insecticide Penncap-M and the pesticide Dicamba. See: <https://en.wikipedia.org/wiki/Monsanto_legal_cases> (accessed 22 October 2020). Of course there are several other important lawsuits Monsanto lost, to do with bribery and waste-site dumping in Wales of Agent Orange. Most of these cases are listed in the above website reference.

15 See: <https://www.bayer.com/en/glyphosate/about-glyphosate-based-herbicides-and-their-role-agriculture> (accessed 22 October 2020).

16 For a mapping of this tentacular reach see: <https://civileats.com/2019/01/11/the-sobering-details-behind-the-latest-seed-monopoly-chart/> (accessed 19 November 2020).

17 I recommend reading this article which gives you a very clear picture on how the 'seed game' is played by these multinationals and how resistances can be formed: <https://www.grain.org/article/entries/5142-seed-laws-that-criminalise-farmers-resistance-and-fightback> (accessed 19 November 2020).

18 Kissinger quoted, in: <https://theecologist.org/2016/oct/10/first-agent-orange-now-roundup-whats-monsanto-vietnam-ecologist-special-investigation> (accessed 22 October 2020).

19 And I suspect that most of you, by now, are aware of Bayer's role during World War Two in relation to the Holocaust. In 1925, Bayer became part of the IG Farben general conglomerate of chemical industrialists. Bayer used slave labour in factories it had built in concentration camps (most notably in Auschwitz III, Monowitz Concentration Camp). It conducted medical experiments in Auschwitz and Malthausen using Jewish women inmates (bought at 170 Reichmarks per person). Post-war, IG Farben's director at Auschwitz, Fritz ter Meer, was appointed as chair to the Bayer supervisory board (1956–64). A public apology by Bayer was not made until 1995 (see: <https://en.wikipedia.org/wiki/Bayer> accessed 31 October 2020). In the more recent history of the dangers of ammonium nitrate, we should recall the explosion in the Port of Beirut, Lebanon (4 August 2020). At first it was declared that this shipment was of fertilisers (in itself very hazardous). However, according to *The Guardian* it is believed that the detonation may have been 'a byproduct of Syrian officials' attempts to source nitrate to use in weapons'. See: 'Explosive Stockpile in Beirut "Linked to Assad Regime" Film Finds', *The Guardian*, 16 January 2021, 39.

20 See: <https://www.statista.com/statistics/271897/leading-countries-by-acreage-of-genetically-modified-crops/> (accessed 31 October 2020).

21 See: Global Map of Biotech Crop Countries and Mega-Countries 2016, figure 1, in: <https://geneticliteracyproject.org/gmo-faq/where-are-gmo-crops-and-animals-approved-and-banned/> (accessed 22 October 2020).

22 See: <https://www.planetnatural.com/gmo-eggplant/> (accessed 2 November 2020).

23 It is heartening to read that Mexican farmers have struck back and that in 2016 they scored a major victory when a federal appeals court ruled that genetically modified

corn cannot be grown in Mexico. See: <https://regenerationinternational.org/2016/03/17/court-ruling-a-victory-for-mexico-farmers-and-anti-gmo-activists/> (accessed 20 April 2021).

24 For the full report see: <https://www.piie.com/commentary/testimonies/us-leadership-global-food-crisis> (accessed 4 November 2020).

25 Marie-Monique Robin was awarded the Rachel Carson Prize for her work on this project (both her film and the book of the same title).

26 A recent disturbing report concerning one of the so-called advantages of the UK's departure from the EU is that, thanks to Brexit and the relaxation of many agricultural rules, gene editing of crops and livestock may very soon be in the offing. See: <https://www.theguardian.com/science/2021/jan/07/gene-editing-of-crops-and-livestock-may-soon-be-permitted-in-england> (accessed 7 January 2021).

27 The interview was broadcast on Granada TV's *World in Action* programme (entitled 'Eat up Your Genes') 22 June 1998. For more details see: <https://en.wikipedia.org/wiki/Pusztai_affair> (accessed 6 November 2020).

28 Some major political personages mentioned in this revolving-door practice are Donald Rumsfeld and Clarence Thomas.

29 See: <https://www.marketsandmarkets.com/Market-Reports/glyphosate-market-8522593.html> (accessed 31 October 2020).

30 See: <https://edepot.wur.nl/522852> (accessed 31 October 2020).

31 A recent study appears to confirm this link between glyphosate and non-Hodgkin's lymphoma, see: <https://www.theguardian.com/business/2019/feb/14/weed-killing-products-increase-cancer-risk-of-cancer?CMP=share_btn_link> (accessed 2 November 2020).

32 For a profile on Dr Huber see: <https://www.permaculturenews.org/2014/05/16/dr-don-huber-gmos-glyphosate-threat-humanity/> (accessed 2 November 2020).

33 For a clear picture of this contamination of our food chain see: <https://www.hillandponton.com/weeds-not-worth-killing-with-roundup/> (accessed 19 October 2020).

34 See for example Cornell University's report: <https://phys.org/news/2018-07-genetically-eggplant-bangladesh.html> and <https://news.cornell.edu/stories/2018/08/gmo-eggplant-crop-expands-bangladesh> (both accessed 2 November 2020).

35 For a useful overview see: <https://navdanyainternational.org/publications/glyphosate-bans-restrictions-across-the-world/> and <https://sustainablepulse.com/2019/05/28/glyphosate-herbicides-now-banned-or-restricted-in-17-countries-worldwide-sustainable-pulse-research/#.X569DS2ca3c> (both accessed 20 October 2020).

36 In *Idiot Cycle* we learn that, 'AstraZeneca is the result of a merger between Astra AG (Sweden) and Zeneca Group of the UK (second biggest maker of cancer drugs behind Bristol-Meyers Squibb). "Zeneca is also maker of fungicides and herbicides (including the carcinogen acetochlor) and owns the third-largest source of cancer-causing pollution in the U.S. – a chemical plant in Perry, Ohio. In 1996 this facility emitted 53,000 pounds of recognised carcinogens into the air"'. See Minter Dial's review of *Idiot Cycle*: <https://www.minterdial.com/2011/02/idiot-cycle-by-emmanuelle-schick-garcia-a-review/> (accessed 11 November 2020). Takes a bit of the shine off its COVID vax.

37 See: <https://www.idfa.nl/en/film/15863603-96d4-4748-8a50-c06cee47b1a6/the-idiot-cycle> (accessed 11 November 2020). The Independent Documentary Filmfestival Amsterdam offers a terrific selection of documentaries, often accessible for free.

38 See: <https://www.healthline.com/health/paraquat-poisoning> (accessed 11 November 2020). Though its use is banned in the EU, UK and Switzerland, Paraquat continues to be manufactured for export in Huddersfield to this day. The weedkiller is linked to thousands of deaths globally. It is extremely toxic and causes death from multiple organ failure over days. There is no treatment and effectively no way back

from just one sip. See: <https://www.theguardian.com/environment/2021/mar/24/scientist-tells-of-relief-after-speaking-out-over-weedkiller-fears> (accessed 24 March 2021).

39 See: <https://www3.epa.gov/pesticides/chem_search/reg_actions/reregistration/fs_PC-034805_1-Jul-04.pdf> (accessed 11 November 2020).

40 See: <https://archive.epa.gov/pesticides/reregistration/web/pdf/0186.pdf>; according to the European Chemicals Agency (2020 report), however, it is a 'substance very toxic to aquatic life' <https://echa.europa.eu/substance-information/-/substanceinfo/100.038.873> (both accessed 11 November 2020).

41 You will not be surprised to learn that, a decade after these two eco-docs on the fate of bees, very little has changed. See: <https://www.theguardian.com/environment/2021/apr/01/toxic-impact-of-pesticides-on-bees-has-doubled-study-shows> (accessed 1 April 2021).

42 China is now (since 2017) the world's largest producer of Glyphosate. See: <https://apnews.com/press-release/pr-businesswire/3fb81297b58d4e55b935a364bc745a52> and <https://www.prnewswire.com/news-releases/research-report-on-chinas-glyphosate-industry-2018-2022-300689149.html> (both accessed 11 November 2020). In recent years China's Glyphosate production is double that of Monsanto.

43 China is the biggest importer of GM soya (80% of the global market, most of which goes to animal feed). See: <https://phys.org/news/2013-09-china-hunger-soya-animal-vegetable.html> (accessed 16 November 2020).

44 Bruno Latour interviewed on France-Inter's *Journal7/9*, 'L'invité de 8h20', 7 February 2020.

45 See: <http://www.fao.org/family-farming/detail/en/c/1277661/> (accessed 21 April 2021).

46 See: <http://www.fao.org/3/ca7036en/ca7036en.pdf> (accessed 17 November 2020).

47 According to the Australian Governmental Agency for Agriculture 9434 gigalitres per annum. See: <https://www.agriculture.gov.au/abares/publications/insights/snapshot-of-australian-agriculture-2020> (accessed 23 November 2020).

48 See: <https://www.nature.com/articles/s41545-019-0039-9> (accessed 21 April 2021).

49 To see the film <https://www.idfa.nl/en/film/163633e6-3e9e-4772-bad3-2d481b4c85ed/lords-of-water/docs-for-sale> (accessed 21 April 2021).

50 See: <https://www.who.int/news/item/15-07-2019-world-hunger-is-still-not-going-down-after-three-years-and-obesity-is-still-growing-un-report> (accessed 18 November 2020).

51 These figures are also clearly set out in the following article: 'Redefining Agricultural Yields: From Tonnes to People Nourished per Hectare' Emily S. Cassidy, Paul C. West, James S. Gerber and Jonathan A. Foley, 1 August 2013 • 2013 IOP Publishing Ltd, <https://iopscience.iop.org/article/10.1088/1748-9326/8/3/034015> (accessed 17 November 2020).

52 This visionary man described himself as a 'comprehensive anticipatory design scientist'. He dedicated his life to making the world work for all of humanity, beginning with the principle of making 'more with less'. See: <https://www.bfi.org/about-fuller> (accessed 23 August 2020).

53 For more information on these two organisations see: <https://www.colibris-lemouvement.org> and <http://www.milpafilms.org> (both accessed 23 November 2020).

54 See: <https://www.permaculturenews.org> and <https://en.wikipedia.org/wiki/Permaculture> (both accessed 3 December 2020).

55 Three hundred scientists recently submitted a report to the UN's FAO (its first on the global state of biodiversity in soils) describing the 'worsening state of soils as at least as important as the climate crisis and destruction of the natural world above ground' adding that 'a large proportion of the Earth's surface has been degraded as result of human activities'. (See: <http://www.fao.org/documents/card/en/c/C

B1928EN/>, accessed 17 December 2020; this report was published on 5 December 2020, World Soil Day.)

56 To understand the enormity of these companies spread in the seed market I draw your attention to the mapping of the seed industry structure, in: <https://civileats.com/2019/01/11/the-sobering-details-behind-the-latest-seed-monopoly-chart/> (accessed 19 November 2020).

57 For a comprehensive read on the history of seed control see: <https://www.centerforfoodsafety.org/files/seed-report-for-print-final_25743.pdf> (accessed 29 November 2020).

58 See: <https://www.grain.org/article/entries/5142-seed-laws-that-criminalise-farmers-resistance-and-fightback> (accessed 19 November 2020).

59 For a fuller picture of this organisation's practice see: <https://www.navdanya.org/site/> (accessed 1 December 2020).

60 See Kokopelli's website: <https://kokopelli-semences.fr/fr/> (accessed 1 December 2020). As Guillet explains, even though he has been taken to court several times by the multinationals, so far they have failed to close down his operation.

61 Only in 2017 did the French government (implicated in the putsch that ended Sankara's presidency) agree to declassify documents surrounding his assassination. The dossier will be examined in 2021 by a military tribunal. See: <https://www.france24.com/fr/afrique/20201016-burkina-faso-le-dossier-de-l-assassinat-de-thomas-sankara-renvoyé-devant-un-tribunal-militaire> (1 December 2020).

62 Primarily manganese ore and gold are Burkina Faso's major exports. Most corporations are government controlled and are unprofitable. On the hunger index, Burkina Faso ranks 65th out of 78 countries. See: <https://en.wikipedia.org/wiki/Burkina_Faso> (accessed 1 December 2020). GM Bt cotton was phased out in 2016.

63 A recent example, in France, of a dairy farmer unleashing himself from being 'imprisoned as a slave to the big companies' (his words) was aired on France-Inter's *Journal 6/9* 'Sortir du système industriel' (6 November 2020). According to this farmer, 80% of the profits go to the big companies, and so the time has come to change things, 'you have to be prepared to go to all ends'.

64 Since the date of this film, urban agriculture has taken off in quite a big way worldwide: from Paris to Singapore, Mumbai to Ottawa, Wellington, NZ, to Shenzen (etc.) and from window-farms and roof-farms to reclaiming vacant lots, from vertical farms to aquaponics and urban forestry – the benefits are considerable on just so many levels.

65 One wonders if the newly appointed (1 March 2021) Director General of WTO, Ngozi Okonjo-Iweala – the first woman and first African to be appointed – will temper the WTO's zeal for open-market principles and turn its attention more toward food security and food sovereignty.

66 See: <https://www.ife.uni-freiburg.de/wachstumswende/interview-nils-aguilar.pdf> (accessed 4 December 2020).

67 See: <https://www.un.org/en/chronicle/article/losing-25000-hunger-every-day> (accessed 8 December 2020).

68 It is hard not to be discouraged when we see that President Modi of India, with his new law to 'liberalise agriculture', is intending to sign deals with Agribusiness multinationals, despite huge demonstrations in Delhi by farmers. See: *The Guardian*, 'Indian Farmers – Protestors Dig in to Fight Modi on New Law' (12 December 2020, 43).

69 See: <https://www.investopedia.com/articles/insurance/090116/5-biggest-microfinance-companies-bbrijk.asp> (accessed 8 December 2020)

70 For more details see: <http://smartcityhub.com/governance-economy/doughnut-cities/> (accessed 9 December 2020).

71 See interview with Raworth, in: <https://www.triodos-im.com/articles/2017/doughnut-economics-an-economic-model-for-the-future> (accessed 9 December 2020). The parentheses are mine.

4

THE EFFECTS OF DISRUPTED ECOSYSTEMS, THE MIGRATORY FLOW AND THE UNSAFE, INHUMANE SPACE OF THE UNHEARD

In the previous chapters we have discussed three of the major contributory causes to the destruction of natural ecosystems (known also as Ecocide) and climate change, beginning with the human impact on the environment thanks to our drive for sources of energy, followed by the polluting damage done to the planet because of the proliferation of human waste and thence to the contamination effect of big agribusiness on soil, water and air. In all three of these chapters, the eco-docs studied have shown how, over the past 200 years, our Anthropocenic activity has assailed the Earth's planetary boundaries, thereby greatly harming the biospheric domain in which we should seek to live alongside all other species. Greed, capitalistic gain, has pushed human endeavour to take away land not just from other human beings (first through colonisation; later, as an effect of globalisation), but also to colonise territories of animal and plant species and other biodiverse ecosystems. Not only have we played an active role in the extinction of numerous species through destroying their habitats and through poaching their land on a massive scale, the net effect of these activities combined with climate change means incurring a loss of up to a third of all animal and plant species by 2070. Already, a million species are at risk because of human activities (that is one-eighth of all animal and plant species on Earth). According to a 2019 UN report 'a quarter of catalogued animal and plant species are already being crowded, eaten or poisoned out of existence' and the

> five direct drivers of change in nature with the largest relative global impacts so far are, in descending order: (1) changes in land and sea use; (2) direct exploitation of organisms; (3) climate change; (4) pollution and (5) invasive alien species.[1]

DOI: 10.4324/9781003130246-4

TABLE 4.1 Table of eco-docs about effects of disrupted ecosystems and global warming

I. Climate change	*II. War/internally displaced persons*	*III. Refugees/immigrants*
1. *The Age of Stupid* UK (2009) 2. *La Glace et le ciel/Ice and the Sky* Fr (2015)	3. *For Sama* UK (2019) 4. *5 Broken Cameras* Fr/Israel/Netherlands/UK/Finland/Switzerland/US/Can/Rep of Korea (2011) 5. *Of Fathers and Sons* Ger/Syria (2017) 6. *Samouni Road* It/Fr (2018) 7. *White Helmets* UK (2016)	8. *Broken Land* Switzerland (2014) 9. *Come un uomo sulla terra/Like a Man on Earth* It (2008) 10. *Eldorado* Switzerland (2018) 11. *Fuocoammare/Fire at Sea* It (2016) 12. *Human Flow* Ger (2017) 13. *Il sangue verde/Green Blood* It (2010) 14. *Sea Sorrow* UK (2017)

The result, says this same report, is that, to date, 75% of the terrestrial environment and 66% of marine environments are severely altered by human actions.

Small wonder, then, that other-than-human species, having been deterritorialised, are increasingly seeking to survive by migrating to territories much closer to our human homes; that sea creatures are moving into new waters – thereby disrupting existing ecosystems because of their own pressing needs. Small wonder also, that human beings, having lost the security of their homeland, their ecosystems, either because of drought, poor harvests, increased desertification of their soil, or because of civil war, border conflicts or other forms of hostilities, find themselves forced to migrate. And it is this enforced human migratory flow which the last set of our eco-docs addresses and which this chapter explores as a conclusion to this difficult trajectory through the tragedy of our Anthropocene Extinction Era.[2] Table 4.1 provides a reminder of the eco-docs in question.

Historic migration has known numerous cycles many of which can be linked back to the interests of colonial powers and their drive to exploit all resources available in order to maintain their economic prowess. These cycles are also induced by a series of famine- or politically-triggered migrations to survive (some of which can be traced back to the impact of colonialism). In the case of colonialism, the most infamous and largest enforced migration in history is the Atlantic slavery migration of Africans to the Americas – some 12.5 million – from the early sixteenth to the mid-nineteenth centuries. The purpose here was to obtain free labour to work for the land-owning colonialists and bring wealth to the European colonising countries (which, in order of volume traded, are Portugal, Britain, Spain, France, the Netherlands, Denmark). Migration during these same centuries also included forcibly dispersing indigenous people by colonising powers (e.g., Indians in the Americas, Aborigines in Australia, Māori in New Zealand). Another enforced migration took the form of deportation to the colonies of criminal elements (most notoriously Britain's transportation of petty-crime convicts to Australia). In the nineteenth century, in an endeavour to stem the impact of Western Europe's tremendous population growth on the nations' sustainability, countries like Britain encouraged the migration of its citizens to its colonies. As to the other main cause of migration, from the mid-nineteenth century until the end of World War One – as a result of famine or limited farming conditions (affecting, amongst others, Ireland, Germany, the Czech Republic, Russia, Italy) or political persecution or religious pogroms (in Austria, Russia and partitioned Poland) – migrations to the USA constituted another large migratory flow (some 39 million). From 1920 until pre-World War Two, refugee figures to the USA were relatively modest partly because of new restrictions, for example the Immigration Act of 1924.[3] Post-World War Two, figures again remained relatively modest with the exception of the impact of the 1948 India/Pakistan partition (incurring 20 million displaced persons).

By 1951, the year of the UN Refugee Convention, figures for forcibly displaced persons were published on a yearly basis by the UN. Until the late 1970s, numbers remained stable averaging under 5 million per year. From 1981 to the mid-1990s numbers began to increase from 10 million to 26 million.[4] Come the twenty-first century, however, figures magnified. To date, 79.5 million people are officially classified as forcibly displaced persons. In terms of gender, of the 79.5 million, 52.5% are male and 47.5% female. In terms of age distribution around 30–34 million are children (that is, 40% are under the age of 18). Figure 4.1 shows the UNHCR data sheet. [5]

Whilst in relation to the global population, this figure of 79.5 million represents, according to UNHCR, just 1% of the global human population, we should take note that numbers have nearly trebled in a mere 30 years (1990–2020). We need also to take note that these figures refer to officially registered or documented forcibly displaced people. For it is estimated that there are between 37 and 70 million so-called 'migrants/illegal immigrants' – that is, undocumented people around the world.[6] In my discussion of the eco-docs, I shall, therefore, be careful to use the nominations of immigrant and refugee. The term '*migrant*' will be italicised, because (as will become clear) it is either used specifically by the UNHCR (to denote a particular category, such as economic *migrant*), or as a general term to indicate the migratory flow, or by politicians, the media and exploiters of cheap labour as a deliberately misappropriated designation.

Presently, to return to the above UNHCR figures, 26 million, one-third, of the forcibly displaced people are recognised as refugees. The 1951 Refugee Convention defines a refugee as follows:

> Article 1 of the Convention defines a refugee as a person who is outside his/her country of nationality or habitual residence; has a well-founded fear of persecution because of his/her race, religion, nationality, membership in a particular social group or political opinion; and is unable or unwilling to avail himself/herself of the protection of that country, or to return there, for fear of persecution.[7]

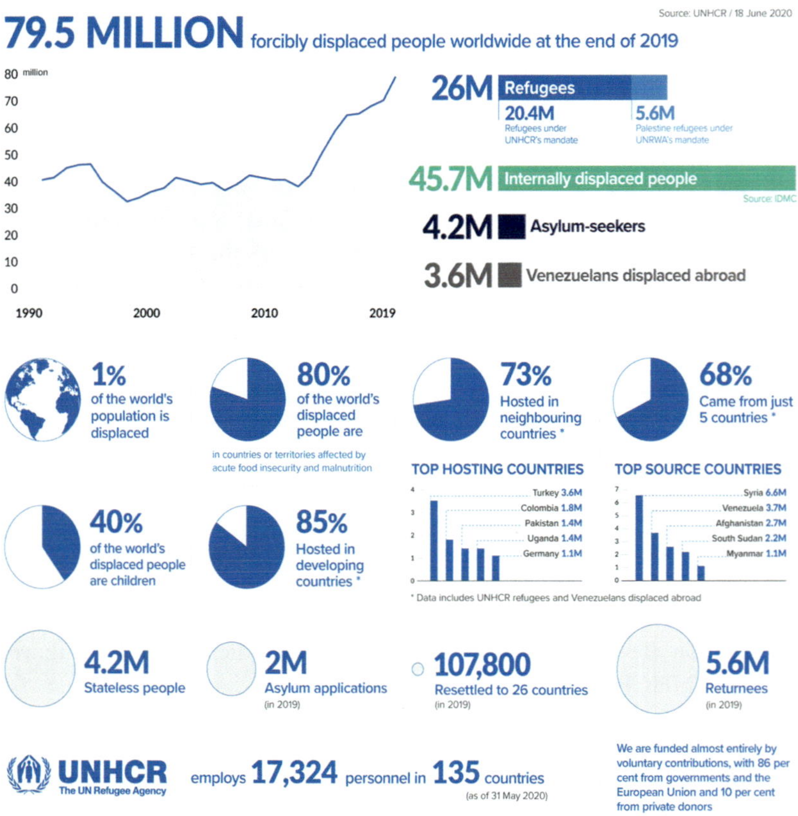

FIGURE 4.1 The UNHCR figures of forcibly displaced people at a glance 2019.

The core principle of the 1951 Convention is '*non-refoulement*', which asserts that a refugee should not be returned to a country where they face serious threats to their life or freedom. Furthermore, Article 31 very specifically states the 'illegal entry' into a country cannot be penalised:

> The Contracting States shall not impose penalties, on account of their illegal entry or presence, on refugees who, coming directly from a territory where their life or freedom was threatened in the sense of Article 1, enter or are present in their territory without authorisation, provided they present themselves without delay to the authorities and show good cause for their illegal entry or presence. The Contracting States shall not apply to the movements of such refugees restrictions other than those which are necessary and such restrictions shall only be applied until their status in the country is regularised or they obtain admission into another country. The

> Contracting States shall allow such refugees a reasonable period and all the necessary facilities to obtain admission into another country.[8]

The Convention further outlines a refugee's rights including such things as freedom of religion and movement, the right to work, education and accessibility to travel documents.

But it also underscores a refugee's obligations to a host government. The Convention does not, however, regulate 'economic' *migrants* who have voluntarily left their country to seek a better life elsewhere. 'Fear of persecution' is, then, the over-riding principle of refugee status.

As for the other sectors of forcibly displaced people, we note that a further 45.7 million are internally displaced people (IDPs). These persons, according to the UNHCR, may have fled their homes for similar reasons as refugees, but they remain within their own territory and, therefore, are still subject to the laws of their own country. But the UN *Guiding Principles on Internal Displacement* (2005) affirm 'that the state has primary responsibility for the rights of IDPs' to protect and assist during displacement and must also permit 'humanitarian organisations to offer assistance' (Ferris and Kirisci, 2016: 74).[99] There is yet another 7.8 million people either seeking asylum or displaced abroad. Finally, the most revealing statistic of all (in Figure 4.1) is that 80% of the world's 79.5 million displaced people are in countries affected by acute food insecurity and malnutrition. In actual figures that means 63.6 million people.

The UNHCR figures above further indicate that 85% of refugees (21 million) are being hosted in developing countries, even countries on the brink of economic collapse such as Lebanon with 1.4 million (which the UNHCR, in this particular document above, fails to mention on its list of top hosting countries; as indeed the 6.6 million refugees in sub-Saharan African nations, of which Uganda has 1.4 million[10]). These are nations without the real means to support this migratory inflow. Yet Europe and other wealthy countries have been less than welcoming (Germany excepted). Shamingly, as several of the eco-docs point out (*Como un uomo sulla terra*, *Fuocoammare* and *Human Flow*), the EU and in particular Italy, since the early 2000s, were making deals with Libya to stem the 'influx' of people seeking refuge from Africa. *Como un uomo sulla terra* even includes footage of Berlusconi and Gaddafi signing a 2008 Treaty agreement (of which I'll say more later when discussing the film). Significantly, Libya is a key transit country for north and sub-Saharan African refugees trying to get to Europe. Moreover, these agreements are signed in the knowledge that Libya is the only North African country not to be a signatory to the 1951 Convention on Refugees nor the 1967 Protocol (extending the 1951 Convention) and has no asylum system – therefore, all immigrants into Libya are considered illegal. More recently, as we know, faced with the Syrian refugee crisis, the EU signed, in 2016, a deal with Turkey – a country which, although a signatory to the 1951 Convention and 1967 Protocol, denies refugee status to non-Europeans (that is, it only recognises European asylum seekers fleeing persecution *in* Europe as refugees). This means that, in both Turkey and Libya, forcibly displaced persons (from Syria and the African continent) have no legal or human rights because they are not recognised as refugees and so can be designated as 'illegal immigrants' and summarily dealt with – which can mean being repatriated (*refoulement*). This lack of recognition of refugee status by Turkey and Libya provides a useful loophole for EU countries and allows them to refer to immigrants as 'illegals' or 'clandestines', as indeed Italy does in its most recent 2017 *Memorandum of Understanding* signed with Libya, thereby justifying their refusal to accept immigrants into their territories – as long as they remain in Libya, Article 31of the UN Refugee Convention does not apply. As legal expert, Anja Palm of the Instituto Affari Internazionali explains, the choice of vocabulary in the *Memorandum* is not innocent:

> This could be interpreted as a (not so) hidden policy choice which aims at the creation of a rhetoric based on the association of different legal statuses to a single category of persons not having the right to enter the EU territory. Indeed, the outsourcing of migration management as a whole necessarily also implies the outsourcing of asylum management, and this in a country (Libya) that does not recognise the right to asylum, nor has the institutional capacity and experience to process international protection claims.[11]
>
> *(my parentheses)*

Billions of euros have gone into sealing deals with these two countries well known for their abuse of human rights – examples of which we hear, from witnesses in *Sea Sorrow* and *Il sangue verde* as well as *Come un uomo sulla terra, Fuocoammare* and *Human Flow*. In Ai Weiwei's *Human Flow* (2017) we get to see the Syrian immigrants' camps, which house a mere quarter of a million of the 3.6 million refugees; after which we then visit the ruins or make-shift accommodation in which the majority live (often up to eight in a room).[12] In Vanessa Redgrave's *Sea Sorrow* (2017) we witness the plight in particular of children in refugee camps and other temporary shelters. As Redgrave asserts, 'the conditions breach the 1989 UN Convention on the Rights of the Child'.

There is something morally obscene about the bartering of bodies to secure frontiers or contain borders. E/Insuring a migratory body remains a commodity of exchange amounts to a corporeal dehumanisation – devalued as less than a human being. This is what gets exposed in 7 out of the 14 eco-docs making up the corpus of this chapter (see Table 4.1, column three). As for forcibly displaced persons in their own country – which is the focus of five eco-docs, within the context of the Syrian Civil War and the Israeli-Palestinian conflict (see column two) – the lengths to which a dominant state power will go to dehumanise its targets are revealed as equally abhorrent. Nor should we forget that these migratory issues (internal or otherwise) are frequently legacies of colonial and post-colonial mismanagements. For example, Syria and Palestine, formerly part of the Ottoman Empire and subsequently – after World War One – mandates or protectorates under France and UK respectively, gained 'independence' after World War Two, but not before the French had to be forced out of Syria in 1946 and Britain relinquished its protectorate – in 1948, thus allowing for the partitioning of Palestine into two countries, Israel and Palestine, and leaving them to get on with it.

The legacy, in particular, of late colonialism of the nineteenth century and its pursuit of technological progress and capitalist growth has contributed (as I indicated in Chapter 1 and explored more fully in *Film Ecology*, Hayward, 2020) to the planet's degradation. In reducing the planet's resources to a commodity-exchange for profit, an economic practice still in place today, we have created a world that can no longer sustain its inhabitants. The previous chapters have provided stark examples as to how we have arrived at this critical point. And by using Raworth's Doughnut model to illustrate the effects of this slow violence of planetary harm, I have sought in my discussions of the eco-docs to stress their invaluable performative service in exposing the damage done – in most instances through localised examples and testimonies, thereby giving voiced and visual evidence of what is at stake (rather than documenting through statistics, as we saw with *Home*, the effect of which can be alienating, or removing the 'human-link'). By plundering the Earth rather than nurturing, by seeing land as a battle terrain (terms such 'taming nature' spring to mind), by treating territories as spaces to be occupied and fought over for ideological or religious reasons as much as for profit, we have managed to create a further contributory cause to planetary harm – this time affecting the human species in the form of the migratory flow.

Central also to this story of the unhealthy economic practice of growth capitalism is the current neoliberal economic model which, since the 1980s, has both constructed and promoted world-wide the economic imperative of globalisation; for it, too, by impacting on the planet and humans, has contributed to the migratory flow. Globalisation, under the guise of spreading Western democracy world-wide, has done more to serve corporate interests – as the previous chapters have amply illustrated (natural resources of the poor nations taken over by the wealthy, imposing mass agricultural processes, pollution of wealthy nations dumped upon or affecting poorer countries and so on). Globalisation, in short, welcomes expansion, often under the premise of creating 'opportunities' for poorer nations; it is not, however, an economic model that welcomes reciprocity, in the form of immigrants and refugees coming into the domain of Western capitalism.

As the above makes clear, many elements contribute to the shattering of eco-systems which then render people either internally displaced or forced to abandon their country – and the eco-docs under consideration in this chapter do much to bring these to light. In the discussions that follow, I need, first, to contextualise these documented stories and their unfolding, after which I shall draw on Raworth's Doughnut model to illustrate the impact of these violations on the planetary and social boundaries and, too, record the few instances when the safe and just space for humanity is attained. We begin the five eco-docs in column two that deal with the consequences of war-torn Syria and Palestine on civilians and the intentional creation of IDPs. After that, the focus will be on the seven films listed in column three, which showcase the plight of refugees and immigrants. In both categories of eco-docs, what prevails are individual attestations to the abuse of human rights, testimonies from participants who have lived the effects of deterritorialisation – that is, the voices of the unheard. The chapter ends with a mapping onto the Doughnut of the documentaries discussed, accompanied by a brief summary of the conditions of shortfall that are exposed (see Figure 4.8).

As to the two eco-docs in column one, *The Age of Stupid* and *La Glace et le ciel*, because they deal specifically with climate change and provide a grand summum of our Anthropocene era, it seems fitting that I should reserve my comments about them for the Conclusion to this book.

4.1 Uncivil wars – disrupting ecosystems from within: the fate of IDPs: *'Indiscriminate attacks on civilians and civilian structures that do not discriminate between civilian and military targets are illegal under international law'*[13]

4.1.1 Syria

> Millions of people watch, no one does anything.
>
> Waed Al-Kateab, journalist-filmmaker (*For Sama*)

First, let us consider the three eco-docs which bear witness to the ongoing Civil War in Syria: *White Helmets* (2016), *Of Fathers and Sons* (2017) and *For Sama* (2019). They make for a powerful triangulation of witness documents for what is occurring in the ongoing Syrian Civil War which began as a consequence of Assad's brutal and repressive response to the Syrian population's own uprising against its government and president during the Arab Spring Revolution of 2011. But this war has to be seen in the light of what had preceded. Many factors contributed to the 2011 out-swelling of grievances, the very first being continuous social and economic injustices brought about in the wake of Assad's Liberalisation Campaign (initiated in 2001). This economic reform involved a shrinking of state involvement in the economy and the building of a market economy, gradually allowing the private sector to take the initiative – in short, entering a market economy by adopting the neoliberal globalisation model. Assad's reforms advocated labour market liberalisation and, to that effect, he signed a number of international agreements on labour legislation. The overall effect was to create a market imbalance in favour of employers, which the regulatory bodies (state institutions) were too weak to compensate for. Given high unemployment levels, employers were able to lower workers' wages and privileges. The result: the gap between rich and poor increased and social differences started to become more apparent. The resentment of the poor could be clearly heard, but only in private since public demonstrations are banned in the country. However, the 2005 United Nations Development Plan (UNDP) poverty report on Syria showed that in 2003–4 around 2 million people (11% of the population) lacked access to basic food and non-food requirements and that the overall proportion of people in poverty was at 30%, including 5.3 million persons.[14] UNDP estimated that, by 2008, poverty rates would undoubtedly be higher still. They were correct. But even they could not imagine the scale of increase, post-2011, as a direct outcome of the Civil War. According to a recent March 2019 report, 83% of Syrians living inside Syria are below the poverty line, the great majority being IDPs whose number, since 2013, has remained constant at around 6.5 million.[15]

Prior to the war, one of the significant Liberalisation Campaign reforms in terms of its impact on migratory patterns was the 2004 Law 56 on agricultural relationships which reduced protection for tenants and laid out new regulations to resolve old rental arrangements. Assad cut agricultural subsidies and privatised state farms. Following the removal of government subsidies, rural Syria was hit by big increases in the price of crucial agricultural inputs. In addition, 2007 saw the implementation of a new agrarian-relations law giving landowners the right to terminate tenant contracts with immediate effect; and 2008 saw the enactment of Presidential Decree No. 49, which restricted land sales (favouring investors) across the whole of Hasakah province in the rural North East – an area known as Syria's Breadbasket because of its wheat production.[16] These major structural reforms – as well as a series of successive droughts (2006–9), plus a late-winter frost destroying their wheat crop (2008), especially in Hasakah province – brought about an internal displacement of some 300,000 persons who migrated from Hasakah to urban areas such as Damascus, Daara, Hama and Halab.[17] These figures are small in comparison to the disastrous situation resulting from the Civil War, but they are indicators of the collapsing of self-sustaining ecosystems induced in part by political factors and, more pressingly, anthropogenic global warming.

Prior to the Civil War, Syria's population was 21 million. Since the war, essentially half of Syria's population are forcibly displaced people: 5.5 million refugees and 6.5 million IDPs. The popular uprising in the 2011 Arab Spring's peaceful protest very quickly turned into a war of assorted rebel forces pitting themselves against the Assad regime. At the beginning of the war there were some 21 rebel factions fighting the Syrian government forces. Most of the territories held by the rebel forces were in the northern provinces: Idlib, Aleppo, Raqqa, Hasakah and Deir ez-Zor. By 2012–13, the point at which our eco-docs begin their documenting, the three dominant rebel forces, ISIS, ISIL (DAESH) and Al-Nusra, held all five provinces. Al-Nusra and ISIS controlled Idlib, Aleppo, and Raqqa; Al-Nusra controlled Deir ez-Zor; and ISIL, Hasakah. As time elapsed Al-Nusra (part of Al-Qaida) became the dominant force, particularly in Idlib (as we see in *Of Fathers and Sons*).

Henceforth, from 2013 to 2017, the time-lapse of our eco-docs, two of the worst affected areas, in terms of warfare and IDPs, became Idlib province and the city of Aleppo, the events of which these eco-docs record. Orlando Von Einsiendel's *White Helmets* follows, in graphic close-up, the harrowing daily rescue routines in the city of Aleppo of the volunteer Syrian Defence Force known as White Helmets. In *For Sama*, Waed Al-Kateab records, for her new-born daughter Sama, the 2012–16 events in the city of Aleppo. As a citizen and mother, she diarises, with her camera, the repeated air strikes against the city's inhabitants and in particular the bombings of hospital and medical centres. Talal Derki's *Of Fathers and Sons* provides us with an insight into the consciousness of Salafi-Jihadism as embodied by one of the founder members of Al-Nusra, Abu Osama, whose overriding belief is that it 'will liberate Syria and set up a fair Islamic Caliphate'; and whose unyielding discipline means that his sons will be trained into the Jihadi philosophy and become fighters in turn.

The bravery involved in gathering documentary evidence needs to be explicated here. All three filmmakers, to a greater or lesser degree, had to go undercover to bear witness. Al-Kataeb, a journalist, began secretly recording the demonstrations in 2011–12 on her phone. By late 2012, this was so dangerous, she began to film events as they played out in the working and private lives of friends; and, too, she records her own trajectory from single woman to married and pregnant and giving birth to Sama. This time she took different risks in a combination of a fly-on-the-wall style of documentary in the hospital and makeshift medical centres (including distressing images of dead bodies) and a slice-of-life shooting of the everyday, but *ab*normal, way of life to which civilians were reduced (with bombs exploding all around their homes, knowing they could be struck at any moment). The filming did not in any way normalise the dreadful conditions in homes and hospitals (where her husband and other friends worked), but constituted what Al-Kataeb termed 'a way of bearing witness to the truth' and 'providing a memorial to those killed'. So convinced was she that she would die, at one point in the documentary she asks her daughter (in her voice-over narration), 'will you blame me for staying here, or blame me for leaving?' Eventually, during the extreme violence of the Syrian government's retaking Aleppo in 2016 (aided by the Russian air strikes and the use of chemical warfare, in the form of Sarin), she decided to leave with her husband and one-year-old daughter. She also succeeded in smuggling out her documentary footage and came to England where, in collaboration with the filmmaker Edward Watts, she produced what she has called her 'love-letter to Sama', the *For Sama* documentary for Channel 4 TV (2019). The film won the BAFTA for best documentary, was Oscar-nominated and upon theatrical release took $1.4 million at the box office.[18]

Both *For Sama* and *Of Fathers and Sons* offer political insights into the causes for the conflict by narrating it through the personal. *For Sama*'s focus is on civilians, people not affiliated with a rebel group but who actively declare their hopes for a change of regime in the form of peaceful civil resistance. With *Of Fathers and Sons* (Sundance Film Festival awarded and Oscar-nominated in 2018), however, we cross over to other side of the fence, as Syrian filmmaker Talil Derki returns to Syria from Germany (where he lives in exile) and goes undercover posing as a war photographer and Jihadi sympathiser. He infiltrates a rebel Al-Nusra community occupying a village in Idlib, and even more precisely the family of Abu Osama, with whom he lives for two and a half years (successfully, without his cover being blown). Abu Osama, it transpires, is the Al-Nusra group leader, a Salafist-Jihadi, a fervent admirer of the Taliban, and committed to the Mujahideen. One of Abu Osama's first declarations is how proud he of his sons (he has four) two of whom (Osama 13, and Ayman, 12 – named after Al-Qaida leaders) are about join the Al-Nusra military training camp and prepare for battle. He is equally proud of having 'cleared the village of the former Sufi inhabitants', whom he labels contemptuously as 'soft, bench-sitting mystics'. Only Al-Nusra is capable of cleansing Syria of the corruption of Assad's people, 'because it is the pure way', declares Abu Osama, adding 'why bother to join the other groups since Al-Nusra is winning'. But he also assures Derki that he only wishes to 'spread love'. Beyond his love for his sons, there is little evidence of this, however. And even when it comes to his sons, if it is love it is harsh. For his brutality is not just contained at the level of language ('I'll hit you with an electric cable' he warns Osama, 'I'll skin you with my knife', to Ayman), he also metes out rough justice for misdemeanours, such as shaving Ayman's hair off and beating Osama for scrapping with other boys in the compound. He takes his sons out of school because it is mixed and because he wants to educate them according to Shari'ah law and the Caliphate. His own rhetoric encourages his boys to be ruthless. For example, they demand permission to decapitate a small bird they captured. Whilst in captivity it had pecked their little brother's finger. The boys call for revenge by repeating 'slaughter'; their father confirms, yes 'it's Halal to kill a bird, Allah allows it'. Throughout the film, the will of Allah is invoked for every misdeed, victory, act of violence or indeed misfortune. Another father states candidly that 'I choose to buy weapons over food for my kids, out of my loyalty to Allah and the Mujahideen, who'll liberate this country'. Sloganeering is rife: 'let the beheading begin', 'die, you bastards', the fathers cry out.

The children ape their elders, unsurprisingly. Thus, Ayman and Osama, who have watched their father dismantle mines in the field, proceed to make home-made mines and, rather successfully, cause a mini-explosion by jumping on them – a slightly bizarre foreshadowing of what actually happens to their father whose left foot gets blown off one day when he is dismantling a mine (also 'Allah's will', Abu Osama acknowledges). Everyday violence, then, is the only visual and verbal rhetoric in which these young boys swim. At one point, Abu Osama takes them to visit sites of his victory over the Assad enemy, including Ma'arrat al-Nu'man, where he explains exactly how victory was won – much of the town lies before their eyes in ruins. Shortly after this expedition, we watch his two sons and other boys from the village (all around 11–13 years of age) being taken to military camp and put through their paces. It is a brutal regime, the most striking images of which show the boys doing drill exercises as the instructor shoots at their feet or between their legs (when standing at ease) so that they will learn to 'have no fear', for 'today we prepare, tomorrow we fight and kill under Allah's flag'. Osama excels; not so Ayman. As the film ends, Derki, in voice-over, tells us, 'Ayman is pursuing his education and taking care of his young siblings. Osama has been led down the path of death'.

These two eco-docs contrast in tone. On the one hand, there is the desire for life, peace and political change without war as expressed in *For Sama*, and, on the other, as with *Of Fathers and Sons*, the fervent commitment to fight to the death to instal a specific ideological dogma and crush all other opposition – a mirror-image you could say to Assad's own commitment. At a third point of the triangle, in terms of tone, *White Helmets* stands as a testimony to the courage of the Syrian Self Defence in war-riven Aleppo and the humanitarian cause they serve as ordinary citizens. The war of Aleppo (2012–16) was fought between rebel forces on the ground and – opposing them on the ground and by air – the Syrian government forces, subsequently reinforced by the Russian involvement. In response to these ongoing attacks on civilians in Aleppo and in cities elsewhere in Syria, the White Helmets' civil defence force officially formed under that name in 2013, receiving training in Turkey supported by the Mayday Rescue Foundation.[19]

In much the same way that the director of *White Helmets*, Orlando Von Einsiedel, came across the story of the Virunga National Park and decided to pursue that story (as we saw in Chapter 1), here again, he heard about the White Helmets and decided he needed to bear witness and volunteer his services. Also, as with *Virunga*, Von Einsiedel, as a means of ensuring the integrity of his approach, commissioned one of the volunteers, Khaleed Khateeb, to be the film's videographer, using a GoPro camera attached to a helmet as part of the process of gathering evidence (we recall in, Chapter 1, he asked Rodrigue Katembo Mugaruka to do likewise with an undercover camera). Further to Von Einsiedel's concept of safeguarding his ethos that the story comes from within, he chronicled, in the form of solo witness-portraits (framed in single-medium-shot), the testimonies of three other volunteers, Khalid Farah, Abu Omar and Mohammed Farah. A third element of the documentary functioned to record these aforementioned men and other volunteers attending a refresher course at the training camp in Turkey. The documentary cuts between this tripartite composition of witnessing which unfolds according to different rhythms. Thus, the stillness of the witness-portraits gets hurtled aside by the chaotic imagery and rapidity of movement of the rescue footage and, in turn, a relative calm returns with the camaraderie between the men at the training camp shining through in the measured tone of their exchanges and their collaboration with each other as they upskill their expertise.

The tone of each element reflects different aspects of their guiding motto: 'To save a life is to save all of humanity'. In the witness-portraits, Kalid, Omar and Mohammed speak of their willingness 'to sacrifice my soul to save people', of it being their 'humanitarian duty' 'no matter who they are, what side they're on, they need our help and it is our duty to help them'. In fact, Mohammed admits that, to begin with, he fought with the government against the opposition until he realised they were bombing civilians, so he left to join the rescuers. The rescue footage gives a sense of the harrowing nature of their work as well as their courageous refusal to give up searching until they are confident there are no more victims to be found. In a specific instance, they persist for 16 hours in digging through the rubble to rescue an infant who they fear may well be dead. Once they manage to extricate him and discover he is still alive, the outpouring of joy simply overwhelms them. They immediately name him 'Mahmood the Miracle Baby'. Two years later, when the men are at the training camp and distressing news about bombing raids in Aleppo brings the men to the brink of despair for their families back home, suddenly, as they emerge into the outside-canteen, who should be there but their Miracle Baby, a healthy well-loved two-year-old baby Mahmood. 'We are so proud', says Khalid, adding, 'the lesson of Mahmood is never lose hope'.

Already within this eco-doc we get wind of the Syrian government's annoyance (aided and abetted by the Russians) at the White Helmets' rescue service (which at the time of this film's making, 2016, had saved some 58,000 lives and lost 130 of its own from the 3000 men and women volunteers).[20] As Abu Omar informs us 'since the Russian intervention, it's become very difficult; they say they are targeting ISIS, but they're targeting civilians'. These same Russian and Syrian officials have also implied that the White Helmets are linked to ISIS and are, therefore, a legitimate target (several of their centres have been bombed). But the real reason is that Russia and the Assad regime are concerned that filmed evidence from the White Helmets on their rescue missions corroborates testimonies received from civilians by Amnesty International, the Accountability Centre and the Violations Documentation Centre (amongst other humanitarian organisations). The White Helmets' evidence (which includes footage of the chemical weapon attack in Aleppo on civilians, in 2016[21]) will be crucial when Assad's crimes against humanity are prosecuted.[22] One can only surmise that Von Einsiedel's powerful eco-doc, in winning the 2016 best short documentary Oscar, will surely serve to add fuel to the cause of humanitarian justice. The question being 'WHEN?'

The information supplied by these three eco-docs colours in the stark statistics emanating from IGOs such as UNESCO and NGOs like Amnesty, Oxfam and Relief-web (amongst others). Twenty-six percent (5.5 million) of Syria's population (21 million) has fled the country; 40% of the remaining 16 million are IDPs, the majority of whom are located in the high-risk provinces of Aleppo (estimated 2.5 million) and Idlib (estimated 1.6 million). Because the IDPs are constantly on the move within these northern provinces in particular (due to the ongoing airstrikes from the Assad regime and Russia), no figures remain stable. There is also a significant number of IDPs in rural Damascus (estimated 2.05 million). In terms of density of location, however, these three are the main areas.[23] The scenario deepens in its darkness when we learn that 52%

of all IDPs are children. The UNHCR *Guiding Principles on IDPs* states clearly that the state has primary responsibility for protecting the rights of IDPs.[24] Thus Assad's attacks on civilian areas constitute a breach of those principles. However, as Ferris and Kirisci point out: 'there is no legal or policy framework for IDPs in Syria' and furthermore 'the Syrian government does not acknowledge there are IDPs in the country' (2016: 79).

The images in the three eco-docs make clear that, in war-torn Syria, there is virtually no safe and just place for humanity. They also demonstrate how war constitutes part of our Anthropocenic desecration of the Earth, impacting severely upon all ecosystems and thereby the biosphere. In fact, in terms of Raworth's Doughnut, these eco-docs show how virtually all of the 12 dimensions of the social foundation have been transgressed – the exception being *networks* in the form of mobile phones and undercover filming which balance out government propaganda and serve a humanitarian cause (see Figure 4.8 at the end of this chapter). We can see from the ruins in Aleppo city and the towns in Idlib province that the impact on civilian life in the entire country is dire. Indeed, 40% of *housing* has been lost. *Energy* is scarce. *Food* shortages are reaching a critical stage (also put at around 40%) and not helped by the loss of agricultural land given over to relief camps (any more than the loss of agricultural land in the early 2000 discussed above). *Water* resources are scarce and often polluted (from chemical warfare attacks amongst other reasons). *Education* is at a standstill for the IDP children, or very restricted to Shari'ah teaching (as with *Of Fathers and Sons*). *Health* is severely at risk (from the bombing of homes and hospitals, for example), as indeed is *income and work* with 78% of IDPs unemployed.[25] As for the dimensions of *social equity, political voice* and *peace and justice* – all three are so evidently violated, whether this results from Assad's brutality or the Al-Nusra dogma (as seen in *Of Fathers and Sons*). Moreover, this latter's ideological practice equally impinges on *gender equality*: for example, there is a brief schoolroom scene, in which the girls sit separated from the boys, but Abu Osama removes his boys from school because girls are present; we never see his two wives whom he dismissively mentions at one point. There is some irony in that Abu Osama's older sons do the cleaning as well as tending to the younger brothers' needs (as for food, are we to assume the cooking miraculously occurs?). *For Sama* and *White Helmets* redress this inequity, the former through the very evident balance between the genders (in professional terms as much as social), the latter through the inclusion of women volunteers and, more poignantly, the eloquent evocation of Khalid, a builder, about his daughter and his hopes for her future that 'she fulfil her dreams'.

4.1.2 Palestine

> I'll plant everything up again, they'll come and tear it out, I'll plant again.
>
> (a Palestinian farmer speaking out at the end of *Samouni Road* a year after the Gaza War, 2008–9)

Palestine goes back into ancient times and has existed as a political entity, if not as a state, even when it was part of the Ottoman Empire and, later, a British protectorate. Israel did not exist until 1948, after which it declared itself a state and from 1967 onwards expanded its territory to own 77% of the former Palestine. The colossal error of the original decision to deterritorialise the Palestinians through partitioning has meant that, since 1948, a land they considered their home has been reduced to less than a quarter of its original territory and carved up into two separate territories – the Gaza Strip and the West Bank – some 71 miles apart and contained within borders controlled by the Israeli army, but also controlled from within their allocated territory by the Israeli Defense Force (IDF). Even though finally, *in 2012*, Palestine was officially recognised by the UN as a state – voted by 138 out of 193 sovereign states (among nations not signatory to this international recognition are Israel, Western European countries, the North Americas, Australia) – nonetheless, the Palestinian territories still remain occupied by Israel.

I feel I need to start this section on the eco-docs about Palestine by first setting out some maps of the Palestinian Territories (see Figures 4.2– 4.4). It will help clarify the complexity of the Palestinian struggle and the Israeli drive for full control of these territories. The first map is of the Gaza Strip with all the Israeli checkpoints indicated inside the territory and on its border (Figure 4.2).[26] The other two are of the West Bank. One shows the extent of the Israeli occupation of the Palestinian Territory (Figure 4.3).[27] The other pinpoints the Israeli border-crossing points and, inside the West Bank, the IDF checkpoints and security control areas (Figure 4.4).[28]

The violent birth of Israel led to a major displacement of the Arab population. Two-thirds of the (then) 1.4 million Palestinians became IDPs, most moving to the West Bank (700,000) and a smaller number (190,000) to the Gaza Strip (the rest mostly fled to other countries). Displaced from their original homes, the agricultural land the Palestinians had formerly worked became inaccessible.[29] It has been and still is the case that, since 1967, Israel is the occupying power of the Palestinian territories; thus, these so-called Palestinian Territories remain occupied by Israelis either in the form of settlements or an occupying Israeli army or both (which is why, frequently, you hear Palestine referred to as Occupied Palestinian Territory). Israel still controls the coastal and air space of the nominally self-governing Gaza Strip (the implications of which are at the heart of *Samouni Road*). As to the West Bank, there are approximately 650,000 Jewish settlers living illegally on Palestinian land, occupying approximately 42% of the West Bank. Moreover, 85% of the Israeli West Bank Wall is illegally built on Palestinian land (the implications of which we see in *5 Broken Cameras*).[30] Israel also controls electricity and water in both territories which greatly affects the Palestinian population (as we shall see in our two eco-docs). For its part, the Palestinian National Authority (Palestine's government) has been powerless at (or ineffective in) stemming the Israeli occupation of the Palestinian Territories.

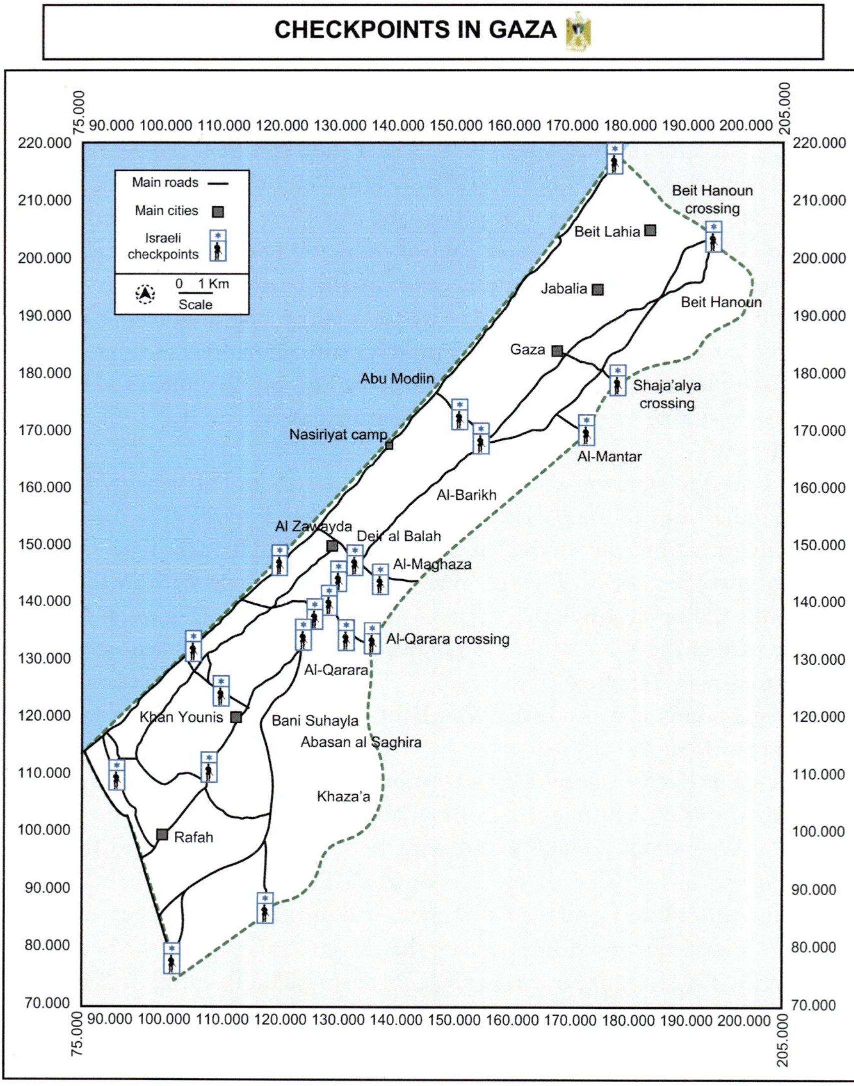

FIGURE 4.2 Israeli checkpoints in Gaza Strip. Zeitoun the location of *Samouni Road* is due south of Gaza City.

FIGURE 4.3 Occupied Palestinian Territory. Area C in light blue is the part of the West Bank under full Israeli control.

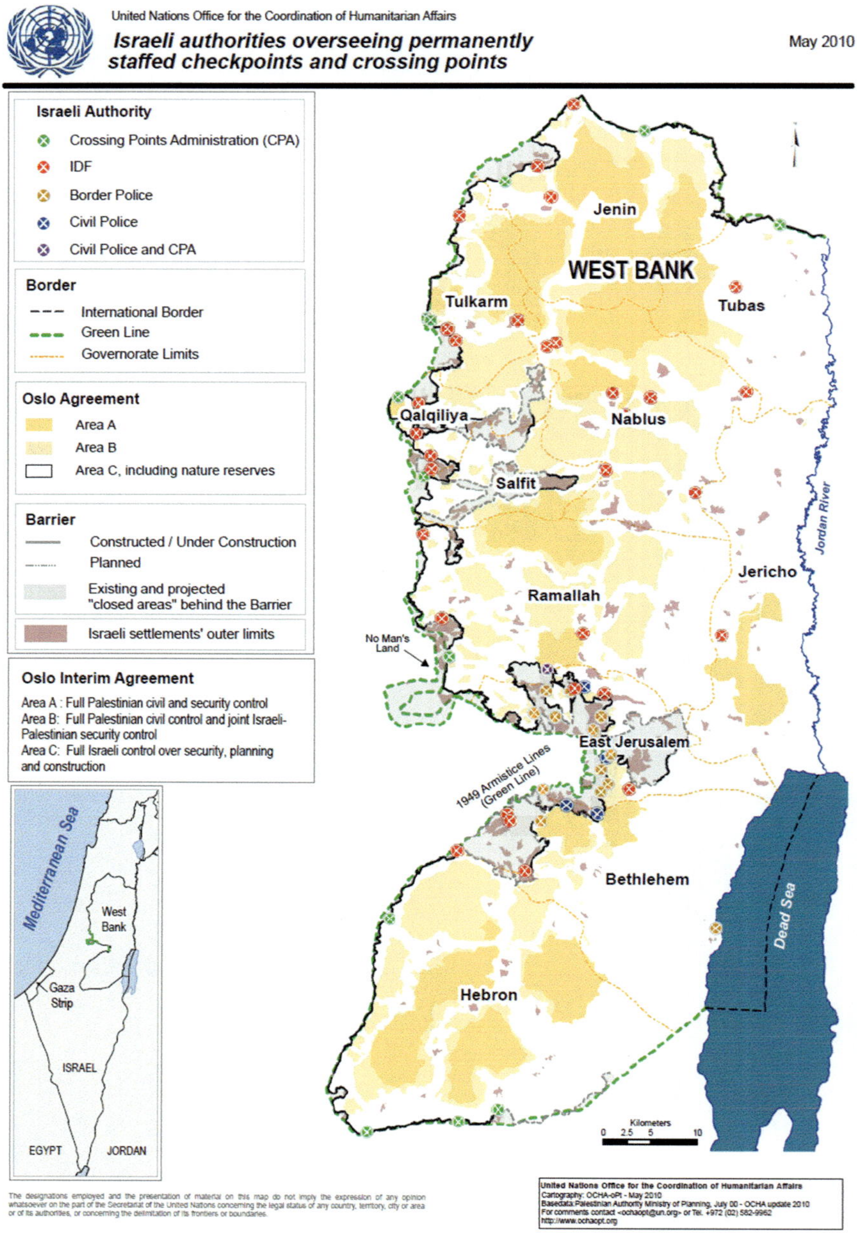

FIGURE 4.4 A map of the West Bank showing Israeli border crossing points, internal checkpoints and security control areas; also shown is No Man's Land.

At this point it needs to be pointed out that as the occupying power, Israel's actions must comply with the provisions of humanitarian law which include observing the Fourth Geneva Convention relative to the Protection of Civilian Persons in Time of War. Amnesty records that:

> Among other things, the rules prohibit the occupying power from wilfully killing, ill-treating or deporting protected persons. According to the Fourth Geneva Convention: As the occupying power, Israel is forbidden from destroying the property of Palestinians in the West Bank and Gaza Strip, unless it is militarily necessary to do so ...: *'extensive destruction and appropriation of property, not justified by military necessity and carried out unlawfully and wantonly'* is a grave breach of the Convention and hence a war crime ... The Fourth Geneva Convention specifically *prohibits collective punishment.*[31]
>
> *(italics mine)*

As we shall see, thanks to the documentary evidence supplied by *5 Broken Cameras* and the witness statements and reconstructions from *Samouni Road*, Israel fails utterly to comply with the Fourth Geneva Convention – more on this below.

Let me now clarify the geography as it concerns *5 Broken Cameras*. Bil'in, the village at the centre of this eco-doc, is due west from Ramallah close to the 'No Man's Land' area and near the red IDF checkpoint on the map (see Figure 4.4), as is Ni'lin (which also figures in the film) although slightly more north-west of Bil'in. Both villages are in Area B, that is, according to the laws of governance, under full Palestinian control and joint Israeli-Palestinian security control. You will note that they are close to the Israeli settlements' pink area immediately adjoining the 'No Man's Land'. This is the territory we see under dispute in *5 Broken Cameras*.

Figure 4.5 provides a further map to show more clearly the locations of these two villages.[32] The red line indicates the West Bank barrier fence under construction (2005–10) which is contested by the Palestinian villagers because it is built, illegally, on their land and separates them from the cultivated land they own (highlighted in yellow), which means they cannot gain access. This lack of access has grave consequences. For, if the villagers are unable to work their land, then Israel can claim their unworked farmland as state land – the effect of which we can see in the form of the pale-blue area of the settlements that constitutes land the Palestinians formerly owned.[33] This law of appropriation is based on an ancient Ottoman law authorising the confiscation of land not cultivated. A number of strategies are used by Israel to provide evidence of non-usage, several of which are mentioned or shown in *5 Broken Cameras*: one, preventing the Palestinians from accessing their land, thanks either to Israeli-built roads to serve settlements upon which Palestinians may not drive (thereby cutting Palestinians off from their land), road blocks, checkpoints, military posts of control or barbed-wire barriers and chain-linked fences; two, Israeli surveillance cameras taking photographs of the land in the winter

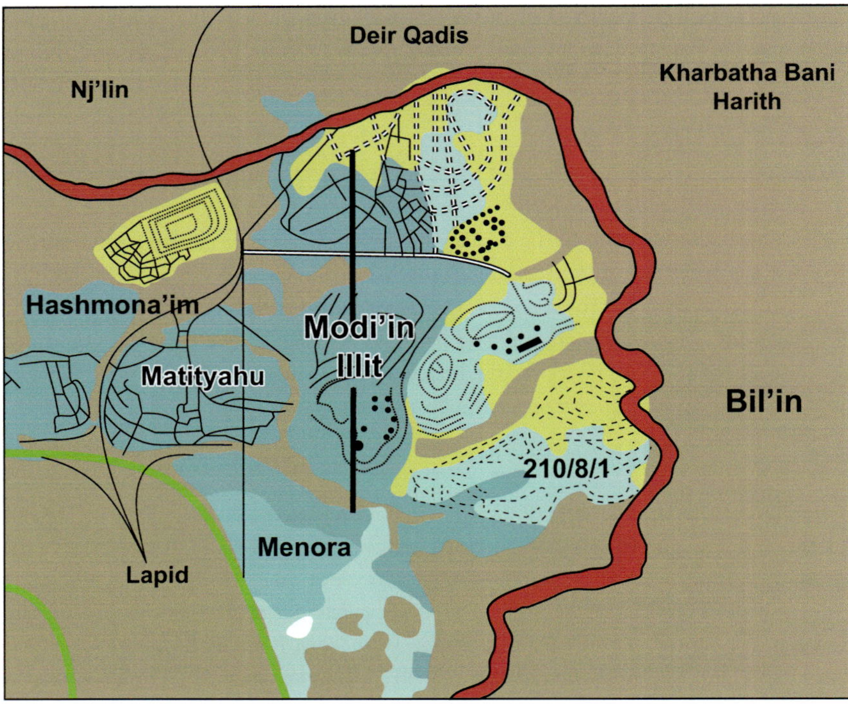

FIGURE 4.5 Map showing Bil'in and Ni'lin. The outlines in green indicate the 'No Man's Land Area'. In red: the separation barrier. In pale blue: the existing Modi'in Illit (Israeli settlements) bloc's jurisdictional area, built on formerly owned Palestinian land. In yellow: the areas included in the Israeli master-plan for further settlements that lie outside the jurisdictional area.

when lying fallow (to prove non-usage); three, developers placing trailers on Palestinian land and immediately claiming occupation. And so on. The other strategy of expropriation is to invoke the plea of military necessity and claim territories have been seized in self-defence.[34]

This difficult tale of dispossession and resistance is recorded by Bil'in villager Emad Burnat, a Palestinian farmer and filmmaker. He got his first camera when his son, Gibreel, was born in 2005, the year the struggle started in his village. 'So, I decided to take part in this struggle with my camera', he tells us at the beginning of the documentary and joins the weekly non-violent protests from the villagers against the armed Israeli military force. 'I know I have a role in life, I know the price may be high, but it is my destiny,' he affirms. Alongside the witnessing eye of the camera, is the concurrent growing up of his child – who becomes an increasingly more knowing eye as he ages from infancy to four years old and whose first words, when two, are 'wall' and 'army'. By three, he goes on the protest marches with his father and recounts the events to his mother, Soraya.

By four, he is asking his father why the soldiers shot dead his friend Phil (the most pacifist of all the protestors and the one whose hope for a better future inspired the village children).

There are five cycles of camera-recording (2005; 2006–7; 2007–8; 2008–9; 2009–10) each one brought about by a breaking of a camera; each cycle witnessing an increase in Israeli military violence. Emad acknowledges 'clinging to non-violent ideas is not easy when death is all around'. Each of these cameras, Emad informs us, is 'attached to an episode of my life inside the occupied territories of the West Bank'. For ease of reference I have sketched out these episodes, in Table 4.2, as a summary-guide to the events we witness. Each column records the conditions Emad and his camera encounter: the Israeli tactics, the Palestinian tactics and the outcomes.

Emad uses his camera 'as witness, as protection and to remember', using the footage and the personal story to expose the cycle of violent response to non-violent protest and to record the resistance against the Israeli illegal territorialising practices of Palestinian land. The film concludes with the information from Emad that,

> after five years of pressure, the court's decision begins to be implemented; the barrier is dismantled; it's a victory, however small, for the village; a concrete wall is being built further from our village; some lands will be returned to the owners, but Bil'in continues to resist the new wall.

And Emad continues to film: 'forgotten wounds can't be healed', he declares, 'so I film to heal, it helps me to survive'.

The appropriation of Palestinian land has meant the loss, by some 60%, of a primary source of its income (olive trees). As significantly, this stripping the villagers of their land constitutes an attack on their identity and generational history. Emad states this quite clearly: 'in the film I say that I am a farmer, but for me every Palestinian is a farmer. The doctor is a farmer, the journalist a farmer, the lawyer a farmer. This is a symbol in Palestine'. Bil'in and Ni'lin are centuries-old villages, the olive trees are generational (some over 100 years old), the farming families are generational. Palestinians are predominantly an agricultural community, yet they have systematically been deprived of much of their land. Furthermore, the continuing Israeli control and exploitation (since 1976) of Palestinian natural freshwater resources in the West Bank, as well as severe restrictions on access to water for Palestinians themselves, have meant that many farmers have had to abandon their land, the impact of which is that Palestinian agriculture is only working at a quarter of its potential.[35] All of the above constitutes a breach of basic human rights (to food and water) which the United Nations General Assembly Economic and Social Council has readily recorded, including the impact on other Palestinian human rights.[36]

In terms of Raworth's Doughnut we see visibilised, in *5 Broken Cameras*, the Israeli processes of appropriation of thousands of Palestinians' farmlands, thereby displacing them and prohibiting access to their source of income. These practices mark a transgression of two of the planetary boundaries – *land conversion* and

TABLE 4.2 The five episodes of *5 Broken Cameras*

	Camera 1 *2005*	Camera 2 *2006–7*	Camera 3 *2007–8*	Camera 4 *2008–9*	Camera 5 *2009–10*
Israeli tactics:	1. Taking Palestinian agricultural land as barrier fence is installed (see red line in Figure 4.5) 2. Bulldozing olive trees (to continue with barrier-building)	1. At the barrier: Israeli military shoot and throw grenades at non-violent protestors (one shot in the leg) 2. Building developers illegally place trailers on Palestinian land to lay claim and extend barrier fence 3. Israeli military continue to harass protestors; prevent them from replanting olive trees on their own land	1. Israeli military illegally enter Bil'in, make arrests including children in attempt to stop demonstrations 2. Emad is arrested (accused of throwing stones); his brother is shot in the leg after arrest 3. Israeli military declare Bil'in a closed military zone	1. Despite legal ruling, fence building continues 2. Israeli military shoot at non-violent protestors killing several 3. Similar shooting at protestors in Ni'lin who have joined the protests (in response to viewing Emad's footage of Bil'in protests, see below); two children killed	1. In contravention of the court ruling, settlers are observed moving into the new buildings ('built on our land' as Emad tells us) 2. Israeli military attack non-violent protestors demonstrating against this occupation and Phil (Emad's close friend) is killed before our eyes by a direct stun-grenade hit 3. At a later demonstration in protest to the killings, Israeli military shoot at protestors, including Emad
Palestinian tactics:	At barrier non-violent protest	1. Attempts to stop illegal trailers being grounded fail, protestors beaten up 2. Protestors decide to place their own trailers on their own land 3. When these are forcibly removed they build a cement shelter which by law cannot be removed	1. In response to the above, children march *en masse* to the barrier fence, Israeli military shoot at them, make arrests 2. Cement shelter still going strong; a lawyer assists in bringing a lawsuit against the barrier-fence construction	1. Realising that morale is getting low, Emad proposes showing his filmed footage to other villages in the neighbourhood, including Ni'lin 2. His footage is now being received by Reuters and outside support is growing	Three separate non-violent demonstrations (see above)
Outcomes:	1. Peaceful protest infiltrated by undercover Israeli police (who cause trouble) allows military to shoot at protestors 2. Stun-grenade smashes Emad's camera	1. Israeli settlers attack protestors for the above tactics; when that fails to deter the protestors, they set fire to a whole hillside of Palestinian-owned olive trees	1. Emad released (case closed) 2. Bil'in re-opened (case closed)	1. Emad tell us 'several villages across West Bank adopt the Bil'in struggle as a model of resistance. The Israeli government fears the protest will become a 3rd Intifada; the military is under pressure'	1. Bil'in resistance 'is attracting a lot of attention from all sorts of political sides' (Emad informs us)

(*Continued*)

TABLE 4.2 (*Continued*)

Camera 1 *2005*	Camera 2 *2006–7*	Camera 3 *2007–8*	Camera 4 *2008–9*	Camera 5 *2009–10*
	2. Israeli settler smashes Emad's camera	3. Warrant issued by Israel's High Court of Justice (HCJ) that finds in Palestinians' favour; orders to stop fence construction and prohibits occupation of the new settler buildings 4. Emad is shot at during protests; the camera takes the bullet and saves his life	2. Israeli HCJ decision instructs that a section of the barrier fence be dismantled 3. A year after the court decision nothing has happened 4. Emad smashes his camera in a tractor accident; he is badly injured	2. Emad's fifth camera is first smashed by an Israeli soldier, mended, then shot at by a soldier in the demonstration protesting against the killings (of Phil and other Ni'lin non-violent protestors)

freshwater withdrawals – which, in turn, entails a transgression of at least 8 out of the 12 dimensions of the social foundation, namely *water, food, health, income and work, peace and justice, political voice, social equity* and *networks* (in the form of service infrastructures). Only at one stage, the small victory of the fence-dismantling, is there a glimmer of a *political voice* being heard and some form of *justice* being exacted in favour of the non-violent Palestinian protestors.[37] To see this illustrated on the Doughnut, I refer you to Figure 4.8.

A similar environment of insecurity prevails in Gaza, the setting for Stefano Savona's *Samouni Road* which relates the impact upon the extended Samouni family of peasant farmers of the 2008–9 Israeli Operation Cast Lead military campaign against the Hamas paramilitary. The Samouni live in Zei'toun, south of Gaza City, and due west of the Al-Mantar IDF checkpoint (see Figure 4.2). In the Press Book[38] for this film, Savona tells us he was in Gaza during the campaign and produced a documentary, *Cast Lead* (2009), recording 'a chronicle of the everyday life during the attack'. Upon the Israeli retreat in late January 2009, he encountered the Samouni family who 'until then had been spared from the sixty years of conflict and occupation and who were facing for the very first time a tragedy without precedent'. A year later, Savona returned at the Samouni's invitation to attend a family wedding – and it was then that he began to piece together the moving story of *Samouni Road*.

The trigger for the Israeli Cast Lead campaign was the increased sense of uncertainty of the Palestinian-Israeli relations, beginning with the Hamas electoral victory in 2006 which removed the Fatah-ruled Palestinian National Authority, the only political party recognised by the Israeli government (and the international community). The escalating violence between Hamas and Fatah forced Palestine's President Mahmood Abbas to dissolve the Hamas-led government and declare a state of emergency. Given the continued unrest, in 2007, Israel imposed a blockade on Gaza borders which meant that Palestinians could no longer travel to work in Israel, nor visit relatives in the West Bank. Israel also instituted internal restrictions which severely reduced the agricultural potential, since over a third of the farmlands were within the restriction areas (see Figure 4.2 for an overview of the control points). Farmlands close to the border were aerially sprayed with herbicides by the Israelis. The impoverished quality of water resources (deemed 96% unfit by the UN) – when added to the situation of food insecurity caused by the above measures – meant a further increase, still, in poverty and social misery amongst the Gazan inhabitants (with 56% living below the poverty line).[39] According to the International Court of Justice and the UN Human Rights Committee, Israel, as the occupying power, was under obligation to ensure the population of Gaza had adequate access to food, essential supplies, medicine and medical care. The prolonged blockade of Gaza, which had already been in place for some 18 months before Operation Cast Lead began, amounted to collective punishment of its entire population which constitutes a breach of the Fourth Geneva Convention.[40]

Following the blockade, rocket attacks from both sides became commonplace and after a brief six-month cease-fire, hostilities resumed in December 2008. Operation Cast Lead, a combination of airstrikes, naval attacks and ground invasion, began 27 December ending on 18 January 2009. The IDF claimed they targeted only Hamas paramilitary installations. The evidence is however that civilians were also heavily targeted (in contravention of the Geneva Convention). The conflict left 1400 Palestinians dead (mostly women and children, 929) and 14 Israeli military (4 shot in friendly fire).[41] On 4 January, Zei'toun, located as it is at the narrowest point of the Strip and near to the Israeli border, was subjected to rocket attacks launched from AH-64 Apache helicopters and an IDF-led ground invasion, including tanks. The attack lasted three days; houses were demolished by air-launched missiles and bulldozed by tanks which also uprooted the community's olive trees. Civilians were rounded up and forcibly removed from their homes; several were shot indiscriminately. The Red Cross and rescue teams were prevented from attending the wounded civilians and retrieving possible survivors buried under the rubble for three days. One of them, 11-year-old Amal Samouni, miraculously survived despite shrapnel wounds to the head and, lying buried under the debris, believed dead (she was unearthed thanks to her young cousin, who had taken a stick to beat the bodies in the hopes he might find some people alive). A total of 29 members of the Samouni family were killed, including Amal's father and uncle Waed.

This is the story Savona decided to portray. But his over-riding concern was how? He 'didn't want to simply make another film of denunciation'.[42] The way the Samouni expressed themselves gave him the clue, for the testimonies they provided drew 'the portrait of a specific community with a fascinating history ... telling something that comes from an oral tradition ... expressing a relationship with the world that is at once anchored in reality and filled with imagery'. Savona continues

> I wanted to free the Samouni from the roles assigned most of the time to Palestinians by the media (as terrorists or martyrs). I wanted to make room for and reveal the variety of their lives, their lives as men, women and children.

This recalls the concept of 'forms of living' we encountered in Chapter 2 in relation to the *catadores*. But again, the question was how to reveal these forms of living and to reconstruct into the living present, the survivors' memories of their former lives and the horror of the attack, and make sense? It was Amal's lack of verbal testimony that provided the answer. 'I don't know how to tell stories' she declares as the eco-doc opens. She then leads Savona to the area in the centre of the village where the 150-year-old sycamore tree used to be, before it was bulldozed away by the IDF. She draws a huge circle in the sandy earth to represent the tree and its spiritual significance to the village – always present and now absent, yet still there in the imaginary – and then paces out the protective shade it used to afford. On several other occasions in the film, Amal draws images (in the soil or on dried bits of wood bark) to represent or map what has been rendered absent – a form of ecopoetics whereby she brings loss back into beingness by inscribing it onto the savaged earth and the bark of destroyed trees (nominally, two provender sources). This was the key for Savona as to how to reconstruct memories of the past. Amal became 'the central protagonist who took me everywhere' and her drawing gave him 'the idea of (using) animation' for the recollections of the past – a task rendered possible by Savona's discovery of the animation artist Simone Massi whose scratch-work technique is directly linked to memory.[43] The laborious process of producing 40 minutes of animation for *Samouni Road* took 7.5 years for Massi and his team to complete. Figure 4.6

FIGURE 4.6 Amal.

provides an example of the etching effect of memory; this is Amal in her family's olive, lemon and almond tree farmland.[44]

The eight animated episodes in *Samouni Road* are based on the stories, testimonies and dreams of the Samouni. By being etched onto the film – in black and white – they act in much the same way as memories do: deeply, permanently, even three-dimensionally etched into our consciousness and unconsciousness. The first six episodes of intercalated animation sequences are lyrical, poetic, reflecting the Samouni forms of living. These are brief glimpses, averaging around two or so minutes, some with haunting background sounds such as thunderstorms or the distant sound of military aircraft, hinting that these memories are clouded by what happened later to destroy it all. When it comes to episode 7 the tone darkens. Birds scatter as F-16 Fighting Falcons fly over. Amusing singing tomatoes on the TV advert-slot are swiftly swept aside by Al-Jazeera images of the bombing of Gaza City nearby. But despite the growing menace of attack, the elders still believe they will not be a target, since they have received no indication from the IDF that they are to evacuate. Episode 7 (lasting 4m45s) concludes with Amal's father Ateya and uncle Waed, amongst others, gathered under their sacred meeting-place of the sycamore tree, coming to the decision to stay.

Episode 8 is a complex cross-cutting sequence between animation and drone footage reconstructing the fatal events (both elements are equally weighted at 9m35s). The former provides the Samouni extended family's point of view, the latter, the Israeli surveillance and firing from an airborne point of view (with three layers of voice-over testimonies: from relatives of the Samouni family; members of the International Red Cross; and dialogue of Israeli air command and pilots, reconstituted from an internal Israeli army investigation). The brutality and chaos that occur from the unexpected Israeli attack take the Samouni by surprise. They believed that, as peace-loving people and peasant farmers with no affiliation to any political group and as an extended family that had belonged for generations to this land they cultivated, they would be spared any violence from the IDF. They had not reckoned with the fact that the IDF had secretly selected Zei'toun not just to set an example (and terrorise other Palestinian villagers), but to occupy this natural choke point on the Strip and turn their village into a military base from which to launch their operations. As one of the elders says later after the IDF retreat,

> the Israelis knew this was a safe and calm area, that's why they came here, knowing they'd make many Palestinian victims. By attacking us Samounis they're sending a message to everyone. It's intimidation so that people will flee, seeing what they've done to the Samounis.

As revealed in *Samouni Road*, the generational, the familial, trust in Earth's providence and a sense of belonging to their land are all key elements of the Palestinian ethos, which is why the trees play such a significant role in this eco-doc. It is through them that loss is most vividly experienced by Amal and her brother Mahmoud – these were their father's trees and, before that, their grandparents, stretching back numerous generations. Inter-generational respect is palpable, and so too across genders. Sons respect their mothers and grandmothers and do their bidding (such as caring for the chickens and weeding the vegetables); older sons long to marry even though they are too poor to imagine it as a possibility. Mahmoud's attempts to push his sister around ('because you're not a man') fail repeatedly – as his father counsels him, 'you must learn to speak well' (that is, respectfully).

The blend, in this film, of live footage and animation is orchestrated in an almost seamless way thanks to the linking transitions between these two core elements – the effect could be described as archaeological, a layered embeddedness between the natural world, the ancientness of trees, the generational understandings embodied by the inhabitants (elders and youngers, men and women) of the need to nurture Earth's gifts – that is until the surprise attack blows those structures asunder. Amal and her brothers recall their father, Ateya's wise pronouncements – often derived from the Koran's surahs – as he tried to guide his children (on the importance of rain, of speaking well, of how humans, as with other animal species, are uniquely different yet the same and all connected).

Ateya is present in all eight animated episodes. Perhaps his most profoundly moving evocation comes in the last episode, as the Israeli attack gathers force in his village and in which, a bit later, we will see him shot dead by the IDF. He asks his children 'have you seen how Allah treats the elephant people? He foiled their attack by sending doves in waves to drop lumps of clay on them'. As he speaks, the animation cuts away to show the elephants trampling through the village wrecking buildings, pulling up the ancient sycamore while the birds grab sods of clay and swoop overhead to drop their *undeadly* weaponry onto the marauding elephants, thus chasing them out of the village.

While the repelling of the Israeli elephants did not occur, nonetheless, the revitalising of the Samouni's forms of living, as evidenced through live footage and animation, stands as a marker of resilience, a willingness to remember and renew again. Just as the survivor testimonies Savona gathered during his second visit to the Samouni family provide a clear picture of the destruction of their ecosystem and its traumatic impact, they also attest to their healing process, through their replanting tiny saplings they have grown and nurtured, through the success of Amal's operation to remove the shrapnel, and through the forthcoming wedding of her brother to Waed's daughter. Finally, and more powerful still, is the Samouni's refusal to allow the various Palestinian political factions to appropriate their traumatic experience to their cause. As one of the elders defiantly tells them, 'we are civilians, we take care of our own dead'. Indeed, were Hamas, Fatah or Jihad to claim the dead as their resistance fighters and 'martyrs' this would feed into the Israeli propaganda that the attack on the village was justified, which it so evidently was not.

Until the Gaza War, Savona points out, the Samouni were not refugees, as opposed to the great majority of Palestinians living in Gaza who are (1.4

million out of 2 million). He goes on, 'since 2009, they have, in some respects, become like other Gaza residents, refugees in their own land'.[45] Indeed, on several occasions we are reminded in *Samouni Road* that the Samouni enjoyed good relations with the Israelis. Waed reassures villagers that because he used to work in Israel and also speaks Hebrew they have nothing to fear. When confronted by the IDF, however, his declarations (in Hebrew) as to the Samouni's being simple villagers and his fail-safe argument of his former employment count for nothing and he too perishes. As the occupying power, it was Israel's duty to protect the civilians, and in this they failed. Thus, with regard to this eco-doc and in terms of Raworth's Doughnut (see Figure 4.8), while many of the transgressions of the two biospheric spheres recall those of *5 Broken Cameras*, the initial catalyst differs (land expropriation by stealth versus open warfare). The intention of the IDF was made clear in the graffiti left on the walls of the Samouni houses they occupied during the attack, namely, to eliminate by force the lives as well as the forms of sustenance of a small peasant-farmer village ('Death to Arabs', '1 down 999,999 to go', among others). The impact and intent of the Israeli Cast Lead campaign on the Samouni's village, then, was the complete destruction of their safe and secure space. To the pre-existing 2007 blockade conditions of living experienced by the Samouni of lack of *water, energy* (electricity), *network* infrastructures, *social equity* and *income and work* (the blockade put an end to villagers working in Tel Aviv for Israeli businesses) was now added a destruction of *housing* and *food* resources through bombing, bulldozing, orchard uprooting, crushing and killing of chicken livestock. The *health* impact is primarily registered through Amal's injuries, but we witness also the mental anguish of many of the Samouni family. *Peace and justice* were denied the civilians, to say nothing of their *political voice* which, on the one hand, Palestinian politicians tried to appropriate for themselves and their cause, and, on the other, the Israeli government attempted to render null and void (claiming the attack was for reasons of security). In short, 10 out of the 12 dimensions of the social boundary were transgressed; as were four of the planetary boundaries: *freshwater withdrawals, land conversion, biodiversity loss* and *chemical pollution* (this latter as a result of the chemical residue from white phosphorus shelling).

This collective punishment exacted by Israel was unlawful. In breaching the Fourth Geneva Convention's prohibition on the destruction of civilian property by the occupying power it negated a cornerstone of the laws of war. The UN Goldstone Mission report (September 2009) found that the Israeli operations 'were carefully planned in all their phases as a deliberately disproportionate attack designed to punish, humiliate and terrorise a civilian population'.[46] The Mission also found, in reviewing the above incidents (with regard to the Samouni family, amongst others), that 'in every case the Israeli armed forces had carried out direct intentional strikes against civilians' and, moreover, that the forces' refusal to allow in humanitarian aid and rescue teams, for three days, was yet a further breach of the Convention.[47] These findings notwithstanding, because Palestine, at the time of the conflict, was not an officially recognised state (not until 2012), the International Court of Justice was unable to pursue the Palestinian Authority's request that the war crimes tribunal investigate the 2008–9 Gaza War.[48]

4.2 Disrupted ecosystems – the plight of immigrants and refugees

> In 1989, when the Berlin Wall came down, there were 11 countries around the world with border fences and walls. By 2016, 70 countries had built border fences and walls.
>
> (Ai Weiwei, *Human Flow*, 2017)

Of the many scripted statements Ai Weiwei presents us with in *Human Flow*, this is perhaps the most chilling evidence of the crisis faced by forcibly displaced people – frozen out by the fortress-walls erected by wealthier nations unwilling to abide by their commitment to the UN Convention on Refugees (1951 and 1967). These are nations well aware that Article 31 prevents the penalisation of immigrants/refugees seeking asylum even if they have made an illegal entry – such as paying a people-smuggler. It is virtually impossible to travel by lawful routes which require immigrant/refugees to obtain visas. These nations also know that 'pushbacks' (whether across the Mediterranean or the Channel, or open borders) are illegal and breach international laws on asylum and *refoulement*. Finally, neither the UN Convention nor EU law requires refugees to claim asylum in the first 'safe' country they reach (the Dublin Regulations were intended to guarantee that asylum claims be spread equitably across the EU).[49]

The 1951 Convention, drafted in the wake of the Nazi atrocities of World War Two, was restricted to European refugees and applied to people who had been displaced prior to January 1951. But then, as new refugee conditions arose and as the process of decolonisation began to have its impact on migratory flows, signatories to the 1951 Convention (with a few exceptions) created the 1967 Protocol which removed these two restrictions. However, in the late 1970s, as refugee numbers began to grow in response to post-colonial issues (such as the persecution of 'political dissidents', or minorities) especially in the African continent (as for example in Uganda), so Western countries began, during the 1980s, to close their frontiers, in particular to asylum seekers from African nations. After '9/11' (the 2001 Al-Qaida attacks on American soil), there was a considerable backlash against asylum seekers and refugees, yet the numbers (after a brief decrease for three years) continued to rise, especially since 2012 (in part due to the impact of the Arab Spring of 2011) from 15 million to 26 million.[50]

As a legal tactic to by-pass their commitment to the 1967 Protocol and minimise the number of asylum seekers they might have to process, Western countries became very busy building walls, barbed-wire and other types of barrier-fences to keep would-be refugee-asylum seekers out. And here is how it works. Moira Paz, in her illuminating article 'The Law of Walls', explains:

> Being outside the wall means being beyond the state's human rights-based responsibility. Second, the way human rights enforcement bodies have treated border walls has made them legally permitted and even encouraged their construction. Immigration walls raise a jurisdictional challenge. Human rights law and the national law of many democratic states guarantee individuals that have established territorial presence access to basic human rights. A porous border is thus required by the very concept of universal human rights.
>
> *(2017: 601)*

To claim the need for a visible border as a means of controlling immigration (in much the same way as airports and ports have border controls) is not, then, in itself illegal. So, humanitarian agencies cannot really step in to prevent them being built. Thus, only a porous border makes for the possibility of refugees' accessing a country where they can apply for asylum. A wall, therefore, annuls a refugee's human rights, but technically is legal. Paz elucidates this paradox of this jurisdictional challenge. On the one hand, 'erecting a wall which closes a porous border is a matter of human rights', on the other, constructing 'a wall as an administrative technique for controlling immigration ... is, from a human rights perspective, a non-event' (meaning, it is of no jurisdictional interest). A non-event because the wall, in that it protects the border and keeps refugees at bay,

> changes the legal conversation in a way that is advantageous to the state, for it enables the court to accept the rejection of asylum seekers at the border in a way that the law could not have tolerated had it been more overt.
>
> *(602)*

Despite the unresolvable paradox that the first option is 'politically unsustainable' (the individualist claim cannot be sustained in a world where states control their borders with their walls) and the second is 'morally indefensible' (a statist claim cannot be sustained in a world that recognises human rights), walls continue to be built and human rights enforcement bodies (such as the UNHCR) 'avoid taking a stand by regulating the physical structure of the wall' – meaning, observing where it is placed. Thus, under the pretext that these walls are built on the nation's border, 'border walls (become) legally permissible to some extent' (603).[51] The only domain in which they cannot be deemed legally permissible is if they are erected on non-state land as, for example, is the case of Israel and its wall-building on Palestine land in the West Bank, which was declared illegal by the UN in 2003 (see above discussion of *5 Broken Cameras*, in the previous section).[52]

Other more porous 'borders' also play their role in this practice of exclusion, thus defying the concept of human rights of which Paz speaks and as the illustrative map of what is occurring in Europe makes clear in Figure 4.7.[53]

Let us just consider the European Union (which until 2020 included the UK). In December 2015, the UNHCR declared 2015 the year of Europe's refugee crisis (with people fleeing persecution and war). The media in general spoke of the European *migrant* crisis – using the label most governments elected to employ as a form of 'invalidating' refugee status (as I explained earlier at the beginning of this chapter, 'illegals' is another such term in use). In fact the crisis – in relation to refugees escaping to Europe – had been ongoing since 2012 (a doubling of the number); 2014 marked a further surge in numbers (this increase due to the wars in Syria, Iraq, Afghanistan, Ethiopia and Eritrea primarily). As for maritime control of migration flows, this has been in place as early as 2006 (see the map in Figure 4.7). Increased building of walls and enhanced Schengen border control began in 2014. Since the 2015 refugee crisis, the EU Schengen Area, the Mediterranean Sea and the English Channel, have come to constitute a form of fortress Europe – even though all three borders are in principle porous borders. For instance, despite the fact that, by 2008, the Schengen Agreement had consolidated some 26 European countries into a single jurisdiction with no internal passport controls at their mutual borders, increasingly, controls and in some cases border fences were being and continue to be introduced (as Ai Weiwei shows us in *Human Flow*). Maritime walls have appeared, beginning 2006, in the form of Frontex maritime operations whose main function is border defence and surveillance and, after the Lampedusa 2013 *migrant* shipwreck disaster, the Italian government's short-lived, more humanitarian undertaking *Mare Nostrum* – which, though a defence measure, was also a search and rescue operation reaching as far as the Libyan coast (see list of 'Maritime Walls' in Figure 4.7). Since 2016, however, a new form of control of migratory flows has come into operation, with EU national authorities instituting legal proceedings against NGO search and rescue (SAR) vessels in an effort to prevent 'illegal rescue operations'.[54]

The term 'European *migrant* crisis', however, needs to be seen in a fuller context. For, when we consider the provenance of the immigrants/refugees and their host countries, the numbers reveal how low on the totem-pole European countries score. According to the UNHCR official register of refugees the figures are as follows. Since 2010, 1 million sub-Saharan African asylum seekers have immigrated to Europe, whereas 6.6 million – that is 25% of the world's officially documented refugee population – have found refuge in other African nations (most significantly Uganda).[55] Transcontinental Turkey presently hosts some 3.6 million Syrians; Lebanon and Jordan together, 2.2 million; European countries, 800,000.[56] Afghanistan refugees number 2.7 million, of which Pakistan, though not a signatory to the 1951 or 1967 Convention, has been host to 1.6 million. By 2016, however, it began a programme of *refoulement* – despite the fact that many families have lived in Pakistan since the 1970s, so many generations do not know their

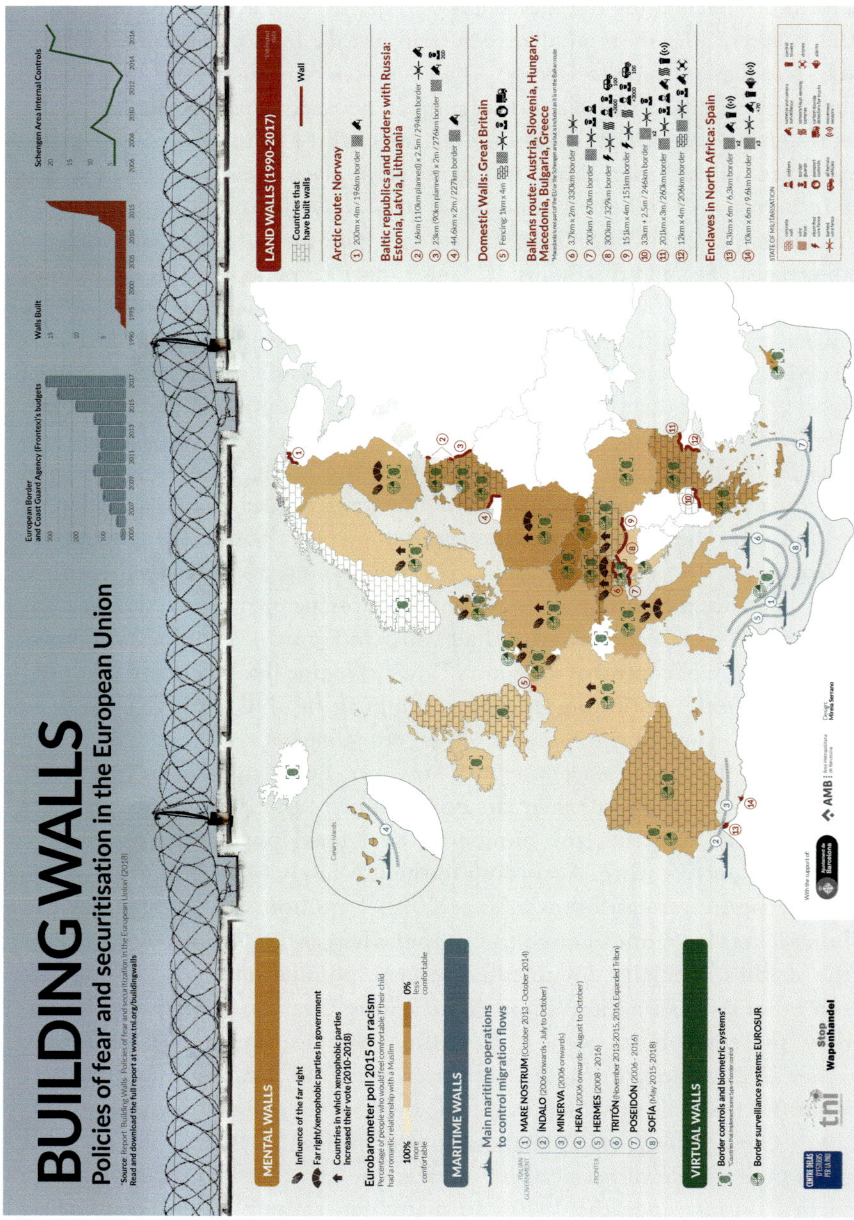

FIGURE 4.7 The European Union's 'wall-building'.

home nation. There are a further 1 million Afghanistan refugees in Iran;[57] for their part, European countries host circa a quarter of a million. Out of the UNHCR official refugee figures for 2019, of the 26 million registered immigrants and refugees, European host-nations account for around 2 million and the USA for 1 million.[58] So, for the year 2019, the Western nations account for 3 million out of 26 million. As to the question of undocumented immigrants, India, the USA and Russia account for the greatest numbers. Estimates put the rest of Europe's intake at around 3–4 million.[59]

This is the world of control to which the six eco-docs in column III bear witness (see Table 4.1). Ai Weiwei, in his documentary *Human Flow* (2017), visits some 23 countries and 40 refugee camps where everything is about barriers, containment and exclusion, and asks the question:

> Can physical borders stop refugees? Instead of building walls, we should look at what is causing people to become refugees and work to solve those conditions to stem the flow at its source. To do so will require the most powerful nations in the world to adjust how they are actively shaping the world, how they are using political and economic ideology – enforced by overwhelming military power – to disrupt entire societies. How do we think the poor, displaced or occupied can exist when their societies are destroyed? Should they simply disappear? Can we recognise that their continued existence is an essential part of our shared humanity? If we fail to recognise this, how can we speak of 'civilised' development?[60]

His eco-doc opens with images of a helicopter flying above an inflatable boat full of immigrants in the Mediterranean Sea near Greece, a sequence which is further developed later on when we see the survivors arriving on the island of Lesbos which is overwhelmed by the numbers. 'Some 56,000 *migrants*, women, children and infants arrive in Greece in one week' we are told by the UNHCR communication officer, Boris Chershirkov, citing conflicts as the source of this migratory flow. For his part, a young Congolese man informs us that he left the Congo because 'when in my country and people spoke of Europe, I thought of Human Rights. But what I see here (as he gestures to the terrible conditions), I don't think Europe exists anymore. This is a living hell'. Many of these refugees make their way – on foot – to northern Greece in an attempt to cross the Greece–Macedonia border and get to northern Europe (we observe them struggling along roads, mountainous passes and crossing rivers, losing possessions along the way). But the border makes any further migration impossible.[61] 'Beginning in 2015, European countries started closing their borders to refugees stranding tens of thousands in Greece', Chershirkov explains, as we again watch the distressing images of these immigrants and their attempts to create some kind of shelter with tents and all sorts of other debris alongside a railway.

Ai Weiwei takes us to visit the desolate conditions in which Syrian refugees find themselves in Turkey – very few of whom are in properly established camps, most of them living in sordid unsanitary accommodation, and who (as I explained in the introduction to this chapter) are all deprived of their legal and human rights. We are also presented with the critical condition of the Rohingya refugees in Bangladesh. Their spokesperson, Ustaz Rafik, speaks of their destroyed life and their humiliation as a people, first, in Myanmar, 'where they have been subjected to repeated campaigns of rape, pillage and persecution', which then continues on another level, in the Bangladesh refugee camp, through a campaign of shaming by labelling them as 'boat people, drifters' and not as worthwhile human beings.[62] As for Pakistan, this country is now deporting all Afghan refugees for 'security reasons'; we see their homes being wrecked, thereby obliging them to 'go back' to a country in which many of them have never lived. Elsewhere, the Serbian–Hungarian border, erected in 2015, serves to defend Hungary from what Prime Minister Viktor Orbán describes as a Muslim invasion that threatens Europe's Christian identity.

The only nations that do not pursue these draconian efforts to exclude immigrants are Lebanon, sub-Saharan nations and Jordan. Jordan, as Princess Dana Firas tells Ai Weiwei, has a long tradition of being 'historically, truly a crossroads for peoples' where

> we have always tried to keep an open door, a place of refuge so that people can retain some sense of dignity and a home until they can return to their homes; ... the health of our own society and relations depends on this.

In Kenya, UNHCR's senior emergency officer Wella Kouyou explains that 'even if there is no violence, climate change will also make people run away; but the resources are diminishing (and) in the next few years climate change will exacerbate drought, hunger and disease for 250 million Africans'. *Human Flow* concludes with a visit to the Mexican–USA barrier wall accompanied by the sobering statement from John F. Kennedy to reflect that 'every American who ever lived, with the exception of one group, was either an immigrant themself or a descendant of immigrants'.

Ai Weiwei describes his net-cast-wide film as a 'personal journey, an attempt to understand the preconditions for humanity in our era' which are 'tolerance, compassion and trust in each other'.[63] He has known first-hand 'how it feels to live in a camp, robbed of (his) humanity'; 'refugees', he asserts, 'must be seen to be an essential part of our shared humanity'.[64] Contrary to the argument voiced by those who wish to resist this human flow by claiming 'that many of these people trying to reach the West are economic *migrants*', Ai Weiwei's amassed visual and verbal evidence says otherwise. The testimonies of immigrants and refugees, the expressions of bewildered fortitude inscribed on the faces of the portraits that punctuate the eco-doc, plus the views expressed by UNHCR personnel among other elements, all attest to the fact, as Ai Weiwei says, 'that these people have fled their homes because they were forced to by violence and that the condition of many of the *migrants* is the direct result of the greed inherent in a global capitalist system'.[65]

This is a view which is strongly expressed, in Gianfranco Rosi's *Fuocammare* (2016), by Dr Pietro Bartolo, the director of health in charge of the incoming refugees onto the Italian island of Lampedusa.[66] To his mind it is clear that these human beings' inability to resource a life in their own country has led to this migratory flow, whether it is a consequence of corrupt practices by governing institutions and exploiting multinationals, civil wars or the effects of land impoverishment. And yet, as Dr Bartolo puts it,

> Europe rejects them, labelling them, categorising them as economic refugees or potential terrorists – as if they would risk life and limb to travel over two years to try and get to Europe – and pays Libya to have them returned and incarcerated.

But, he adds, given 'an ageing continent such as Europe, we could so easily absorb them all' and, moreover, 'we need these young courageous people'.

These are sentiments echoed, incidentally, by some of the US residents living near to the Mexican–USA barrier wall in Stéphanie Barbey and Luc Peter's *Broken Land* (2014) – a film I don't have space to consider, sadly; nor indeed Vanessa Redgrave's *Sea Sorrow* which has much in common, on a smaller scale, with Ai Weiwei's *Human Flow*. I also must leave aside Markus Imhoof's moving personal narrative, *Eldorado* (2018), that intertwines the current experience of many *migrants* and immigrants seeking safety along with his own boyhood memory of a migratory body and her expulsion (a World War Two Italian refugee girl who came to live in Switzerland with his family, but who was abruptly repatriated, after the war, back into poverty and died soon after). There is considerable overlap in *Eldorado* with events and issues raised in *Fuocoammare* and the two other Italian eco-docs, *Come un uomo sulla terra* and *Il sangue verde*, which deal with Italy's particular immigrant-refugee and *migrant* situation and the Libyan-Mediterranean-Italy axis of people-smuggling, abuse and *refoulement*. So, it is to this Italian trilogy that I now turn my attention as a concluding section to this chapter (but I will flag up some of the overlaps).

Shot between 2014 and 2015, *Fuocoammare* covers events that occurred a few months after the 2013 Lampedusa Disaster in which over 300 immigrants perished at sea and in which every islander contributed to help the survivors. It is a documentary that bears witness to the experiences of the immigrants who fled their countries, primarily from the sub-Saharan African continent but also the Middle East, and congregated in Libya whence they attempted by all means possible to cross the Mediterranean, only to fall into distress in the waters between the ports of Tripoli or Zuwara on the Libyan coast and the Italian island of Lampedusa (a distance of some 184 miles[67]). We pick up the story as the survivors are being rescued, then taken by Italian coast guards to Lampedusa and subsequently transferred to the Immigration Reception Centre once they have all been examined by Dr Bartolo.

Historically, Lampedusa is very much a migratory space itself – over the centuries it has been used as a garrison and outpost, then abandoned, later colonised then left unoccupied and finally resettled post-1841 under the Neapolitan flag. Since the mid-1990s, Lampedusa (with its 6000 inhabitants) has become, once more, both garrison and outpost, for it is the primary European entry point for immigrants coming from Libya, as Dr Bartolo tells us. As of 2000, the island morphed into the most visible location of Europe's new border regime. It became both centre (as Europe's visible boundary) and peripheral, as a marginal outpost of Italy; both a place of sanctuary (in terms of the Italian Coast Guard carrying out the rescues and the medical care the immigrants receive) and rejection, in terms of Europe's determination to keep the immigrants out by terminating the maritime rescue operation *Mare Nostrum*, in 2014, and refusing to accept refugees into their countries.[68] As such, Lampedusa finds itself caught in the cross-hairs of UN humanitarian discourses and anti-immigrant sentiments of political parties and populist movements across Europe. Europe, as a continent that has closed its doors as much as possible to the refugee and '*migrant* problem', can hardly be said to embody the inscription on the sculpture of an open door at the southernmost end of the island of Lampedusa. Titled *La Porta d'Europa/The Gateway to Europe* (2008), it is a monument to the immigrants who perished crossing the treacherous stretch of sea between North Africa and Europe – the irony will not fail to be grasped by immigrants/refugees and islanders alike.

The film also bears witness to the inhabitants of Lampedusa: the poverty they experience, the trials of the fishing community and the perils of the sea, especially during bad weather, to which is now added their shared distress at the unceasing traumatic events of refugee boats floundering in the waters and the felt humanitarian imperative to go to the rescue. Hence the significance of the film's title *Fuocoammare* which is also the name of a Sicilian wartime song that refers to the bombing of the Italian naval ship, *Maddalena*, by the Allies in 1943 in the harbour of Lampedusa – for it binds the two concurrent stories (immigrant and islander) together. The text of the original song, *Chi focu a mmari ca c'è stasira/What a Bitter Fire that Was Tonight*, was lost, only this one fragment remaining – much like other transient bodies to touch upon this island – and refers to the visual effect of the 1943 bombing that made the sea appear to be on fire. Song and film title become synonymous with the double truth of the sea and its intimate paradox as a place of nurturing (food from the sea) and danger (drowning, being set on fire), all elements of which affect the islanders and the immigrants. But the word 'bitter' is also key. For, such is the life for the islanders who experience fishing in perilous conditions and the brutal harshness of their work, to say nothing of the hardships entailed if they cannot put to sea. As for the immigrants on this perilous stormy sea, they barely survive – many die either from de-hydration, starvation, drowning or by suffering the effects of skin burns from the fatal mixture of diesel and sea-water down in the hold, where the great majority of women and children are gathered (partly for safety and partly because these are the cheapest fares, at $800). As Dr Bartolo explains, pointing to his photographic evidence, these diesel burns can and do kill.

We learn, from the various immigrants' testimonies, about the terrors of their journeys to Libya (often taking over two years to travel from such countries as Niger, Nigeria, Somalia, Côte d'Ivoire and the Sudan, with many dying en route). We also hear – from the immigrants, as well as Dr Bartolo (who has kept a photographic and written record of all the refugees he has dealt with over the 25 years this migratory flow has been going on) – about the subsequent horrific treatment they are subjected to once rounded up by the Libyan police. One Nigerian sums up their migratory experience:

> we were forced to leave our country and having survived the trauma of trekking all across the Sahara to Libya, Libya had no pity for us because we are Africans. We were put in prison, we were beaten, many died. Then we took to the sea.[69]

A further set of survivors, this time women, remain predominantly mute, crying, their faces torn by grief and loss as they endeavour to comfort each other.[70] Their experience is filled in, later on, by Dr Bartolo, again from his recorded evidence. The women in the Libyan camps or prisons were commodified into different groups. The pretty young ones, he tells us, are set aside to be sold on as prostitutes once they get to Europe (and also forced to take pregnancy-inhibiting drugs that will eventually sterilise them); the less attractive women are used as 'whores' and regularly gang-raped by Libyan soldiers and prisons guards until they 'gain' their passage aboard a boat.

Immigrants relating their stories in *Come un uomo sulla terra* (Andrea Segre and Dagmawi Yimer, 2008), *Il sangue verde* (Andrea Segre, 2010) and *Eldorado* (Markus Imhoof, 2018) provide similar testimonies of the inhumane treatment they received in Libya. But they also make clear that once in Italy, having survived the terrible ordeal of the sea-crossing and rescue, there are further elements of inhumanity to encounter. These three eco-docs along with *Fuocoammare* span a ten-year period of 2008–18 in which nothing has changed in terms of the migratory experience. The small moments of individual kindness along the path of their human flow notwithstanding, for the most part they receive terrible abuse all along the line. These migratory bodies had already been commodified once, in Libya, as sexual commodities of exchange for passage to Europe, or as 'illegal immigrants' to be beaten and tortured, or, if caught trying to flee to Europe, as economic hostages to be sent back to their homeland (by *refoulement*) thanks to a 2008 Treaty signed between Berlusconi and Gaddafi (and reaffirmed in the already mentioned 2017 *Memorandum of Understanding*). Having arrived in Italy, these migratory bodies now find themselves recommodified once again – often at the hands of unscrupulous organised criminal mafia gangs – either as sex slaves or as agricultural serfs (*migrant* labour). Very few move on to better conditions.

Segre's two 'participative' eco-docs delve into this tale of two outcomes. The term 'participative' is Segre's, to refer to his process of engaging with the human profundity of someone who lives and thinks in their particular situation. Thus, *Il sangue verde* is a story of the brutal economic exploitation of African *migrant*-labourers; whereas in *Come un uomo sula terra*, Dag Yimer, co-credited as director of this documentary, recounts his three-year trajectory from Ethiopia, via Libya to a place of relative safety in Rome. Yimer was a law student, but the severe political repression in Ethiopia convinced him to emigrate. His arduous crossing of the desert between Sudan and Libya was nothing compared to the police brutality he encountered in Libya and the violence of the people-smuggling networks in charge of the sea-crossing.[71] Once he made it to Rome, Yimer attended Italian classes where he met fellow immigrants – women and men – and began to accumulate their stories, along with his, with a view to making a documentary about their awful voyage and dreadful treatment in Libya. His intention was to expose the unconscionable collaboration between Berlusconi's government and Gaddafi's regime over immigrants and to provide evidence of the incomprehensible silence about what goes on in North Africa. From there he joined forces with Andrea Segre to make the film.

At the core of this eco-doc is the commodification of the immigrant body into an economic entity of exchange value. And it goes as follows. On 30 August 2008, Berlusconi and Gaddafi signed a Treaty on Friendship, Partnership and Cooperation, thereby putting an end – as we learn from the TV announcer's voice-over footage, watched by Dag, of the two leaders signing the Treaty – 'to forty years of controversy; the colonial contention has been settled with a €5bn deal over twenty years'. The 'controversy' refers to Gaddafi's expulsion, in 1970, of foreign colonial and imperialist influences from Libya shortly after he seized control as the country's president, following the 1969 coup d'état. He expelled the Italian community along with a small community of Libyan Jews and expropriated the Italian-owned assets. With this new Treaty, Italy apologised to the Libyan people for the damages they incurred during the colonialist rule and committed to pay compensation. But the third part of the Treaty, Cooperation, makes clear what was really at stake. 'M. Berlusconi said', the TV announcer continues, 'we can expect greater attention in the fight against the immigration racket, more gas and oil, the opportunity for our companies to increase their turnover'. It escapes none of us that such a treaty – based on the value-exchange of human beings for commodities and profit – resonates with slave-trade deals. To this end, Gaddafi's rhetoric around this part of the deal is significant. He declared: 'with this influx of starving and ignorant Africans, we don't know if Europe will remain an advanced or unified continent'; and he goes on to suggest that Europe could 'turn black' unless it took stringent measures to send 'them' back. This is a sentiment echoed in the eco-doc by Berlusconi's minister for foreign affairs, Franco Frattini, who opines (in the build-up to the signing of the Treaty): 'should the great flux from Libya to Lampedusa continue or resume, then Italy will have to insist on compliance with these international agreements, particularly by means of serious negotiations with Libya'. So, in exchange for Italy's contrition and compensation, Libya commits to keeping the immigration problem under control (keeping the 'black threat' at bay) *and* Italy can obtain more oil and gas from Libya (to feed its 2004 pipeline), *plus* new business ventures – reversing the loss of those former Italian-owned assets. Finally, Berlusconi can remain comforted that, because Libya has not signed up to any human rights conventions, those 'immigrants' under Libyan control will be processed as 'illegals', put in prison and returned home. And all for €5 billion, some deal.[72]

So far in *Come un uomo sula terra*, the immigrant body's exchange-value has been that of a political-economic pawn; but equally, as a number of testimonies inform us, it becomes an instrument of revenue for the Libyan police. As John explains, 'the prison has become a business centre'. When extraditing the immigrant (*refoulement*), a curious circular economy occurs – greatly assisted by container lorries supplied by Italy for deportation purposes. There are as many as '110 *migrants* to a container', John tells Dag, 'the Libyan police then sell us to Sudanese intermediaries, for 30 dinars each'. However, those immigrants who get re-arrested are 'brought back to Tripoli, where the whole process starts up again'. John tells Dag he 'was arrested seven times and sold five times, always for 50 dinars'. A truckload of immigrants means a bounty of around $300–600 for the Libyan police. When measured against a monthly police salary of $99, that is quite a cop. Another source of Libyan police 'revenue-reward', as Mima explains when interviewed by Dag, is the sexual abuse and violence she and most women are subjected to when the police demand money, knowing they have none. Abuse that is compounded, as Mimi and Tighist both relate, by the racism they, as Black African women, have to endure: the police

> gave us food when they wanted, they called us Jews … a policeman poked my pregnant friend in the belly with his baton and said, 'you have a little Jew in your belly, you go to Italy and Israel to fight the Arabs.'

These terms are deliberate attempts to demean the women by insulting their Christian faith as much as their Black-Africanness and, moreover, echo Gaddafi's own racist remarks and hatred of Jews (e.g., their expulsion in 1970).

Racism in all its ugliness raises it head here as an instrumental part of what David Livingstone Smith calls 'cultures of dehumanisation' (2011). To which Dag Yimer responds, 'when asked what ethnicity I am, I reply "I am Ethiopian, period; I am a man"'. A similar declaration of humanity occurs in Segre's other eco-doc, *Il sangue verde*, when one of the protagonists (as Segre calls the seven men whose testimonies he recorded), John says: 'we are all the same, our blood is the same, not black but red, no one has green blood'. As with his previous 'participative' documentary, Segre goes to the source of the unheard – the victims of a culmination of violent racist attacks which took place in Rosarno, Calabria, between 2008 and 2010, with special reference to the events of 7–8 January 2010. Following a number of racist attacks and gang shootings where several African labourers were wounded, they petitioned the local authorities for respect and more humane treatment in a series of peaceful protests. These fell on deaf ears. The labourers eventually reacted on 7 January 2010 when local gangs took pot-shots at them returning from work. Over 2000 Africans took to the streets in protest, where they were met with a combination of local townspeople's hostility and riot police. The ensuing riot was described by the government spokesperson, Berlusconi's interior minister, Roberto Maroni, as evidence of 'excessive tolerance toward illegal immigration' and he ordered the expulsion of every immigrant in Rosarno. The immigrant labour force, some 5000, were immediately rounded up and forcibly 'evacuated' (Maroni's term). The 'problem' was decreed 'resolved' within a matter of hours.

The mediatisation of the event by Italian networks provided none of the above background to the January riots, and was clearly endorsing the government's accusations of dangerous immigrant criminality instilling fear in the inhabitants (a TV camera at one point offers a close-up on a shop-door sign: 'ciuso per paura/closed because of fear'). The television coverage also served the government's consensus-forging among the fearful Italian public by broadcasting politicians' statements to the effect that legality had been restored and that undocumented immigrants would soon be deported from Italy, whereas in fact they were merely dispersed and, indeed, shortly after this mass evacuation, a significant number of the 'evacuated immigrants' returned. And this is the setting against which Segre presents the untold story provided by the seven protagonists – Amadou Bodian, Abraham Kwasi Appiah, Abraham Yabrè, Kalifa Soumahroro, Jamadu Bagayogo, John Kofi Boateng, Tibi Saidou Zongo – and Giuseppe Lavorato (former communist mayor of Rosarno).

The context is yet again linked to economics and the commodification of the migratory labourer – including the exploitative practice of playing on the ambiguity of this labour force's status as either 'migrant' or 'illegal immigrant'. Here goes. Rosarno is an important agricultural centre famous for its citrus groves – where, since the 1990s, the principal migratory labour force has been African (especially for the fruit-picking season). However, ever since the 1970s, the town, with its fertile plain of Rosarno, has been under the control of the local mafia, the Ndrangheta, who have successfully infiltrated all political parties (with the exception of the Communist Party), created 'links' with local businesses and ensured cheap *migrant* labour, thereby weaving a network guaranteeing their interests are served. Thus, the 'illegal immigrants' have been extremely fruitful, economically speaking, for both the Ndrangheta and the citrus grove industry. The same is true of the local mafia-run gangs, politicians and police who have served the Ndrangheta's interests well by suppressing any legitimate demands for fair treatment by the *migrant* labourer who is paid 25 euros a day, provided he is chosen to work the fields, lives in abject squalor, only to be subjected to round-ups and *refoulement* at the end of the orange-harvest season.

The seven protagonists make clear the indecent working conditions and racist treatment they have to endure. They also refer back to the terrible abuse they suffered in Libya – which suggests that the inhumanity dished out to them in both Libya and Italy is but a matter of tiny degrees. Giuseppe Lavorato also speaks out against the treatment the immigrant African labourers receive. Throughout his political career, Lavorato had refused to be intimidated by the Ndrangheta. But, of course, when he retired, in 2003, his anti-mafia political front was swiftly erased by the mafia powers, and the 'swing' to Berlusconi's Forza Republica meant that the new mayor, Carlo Martelli, would present no obstacle to the mafia's control of the production of fruit and vegetables in the region, nor indeed to the continued supply of cheap labour from African 'illegals'.[73] This situation is well known by the TV media and the press, yet – Lavorato's defiant stance excepted – the degree of the local mafia's involvement in the economics and indeed the politics of the region, to say nothing of their brutal control of the entire production supply-chain, is never challenged.

Until, that is, Segre gives voice through his protagonists to the demeaning consequences of a government-sanctioned institutionalised racism which the mafia's mode of operation quite readily enforces. As Alessandra Corrado so succinctly puts it, the 2010 revolt of the African immigrants was 'a reaction against the local racist system, an "ultra-racism" stemming from different factors: the selective and stigmatising differentiation promoted at a political and institutional level; the violence perpetrated by local criminality; and the sociocultural factor of exploitation and exclusion' (2011: 199). Where is the difference, one might ask, between, on the one hand, the Ndrangheta's drive to reproduce itself as a social and economic system in an entire territorial area by 'domination through generalised violence, control and social consensus' (197) and, on the other, a state that endorses institutional racism – and permits its own citizens to 'man-hunt blacks'[74] during the round-ups – as a consequence of its policies on immigration? Policies, I must add, which clearly run contrary to Article 1 of the UN Convention on Refugees, but more significantly, still – given this context of the labelling of immigrants as 'illegals' – contravene Article 31 which states that *illegal immigrants seeking refuge in Europe – and elsewhere – have the right at the very least to present their case and to fair treatment.* All countries that are signatory to the Convention are bound by this agreement and by Articles 32–34 (which include the principles of fair treatment and *non-refoulement*). Alas, we are now in 2021 and the situation is no better than it was then.

The Italian trilogy of eco-docs paints a dismal picture of Europe's callous response to the immigration issue. Apart from Germany, Europe's collective push has been to keep immigrants firmly outside of its territories. Political parties have weaponised the international migratory flow to their own advantage (including posturing about the 'immigration problem' to electors). Yet, at the same time, they are only too ready to turn a blind eye to the exploitation of 'illegal immigrants' as commodities of cheap labour. To quote former Rosarno mayor Carlo Martelli, 'national institutions prefer to ignore the illegal status of *migrant* workers because the already crippled agricultural economy would collapse without them'.[75] This cynical exploitation is, as Kalifa Soumahroro explains, in *Il sangue verde*, 'a new form of slavery'. Indeed, every single article of the Human Rights Act is violated. The precise thing that is not on offer is a safe and just place for humanity where these immigrants are concerned. Their forcible displacement, as a result of broken or destroyed ecosystems, has brought them to an environment where all 12 dimensions of the social foundation – as delineated in Raworth's Doughnut – have been transgressed. It does not make for a pretty picture.

I conclude this chapter with a mapping, onto the now familiar Raworth Doughnut, of the nine eco-docs I've discussed in detail in this chapter (Figure 4.8). The experiences refugees and internally displaced people have been subjected to do not show wealthy Western nations in a good light. Nor indeed can we fail to remark how, whilst the UNHCR does strive to assist these forcibly displaced persons, the UN itself is remarkably without effect in exerting pressure on rich countries to meet their obligations, anymore than it is in imposing effective sanctions on nations that break the UN Convention on Refugees, the UN Universal Human Rights or its own updated 1949 Fourth Geneva Convention Relative to the Protection of Civilian Persons in Times of War.

You will see that, in terms of shortfalls within the Doughnut, *energy*, *water* and *food* are particularly prevalent in the eco-docs about Syria and Palestine – with, in Palestine's case, the impact of transgressions of the planetary boundary on four counts: *biodiversity loss*, *land conversion*, *freshwater withdrawals* and *chemical pollution*. *Gender equality* in shortfall is signposted in at least five eco-docs (especially the treatment of women held in Libyan prisons and camps). *Education* is the lowest in number, but not in significance of course. Indeed, in *Of Fathers and Sons* and *Human Flow*, we witness the deleterious effect, respectively, of a very narrow-band non-secular teaching and the complete lack of any available schooling (contrary to the UN Convention on Refugees). However, the absence of this shortfall, in the seven other eco-docs discussed, indicates something about the implicit level of education of those persons we encounter (for example, Amal in *Samouni Road* continues her schooling, despite the traumatic effects of the Gaza War, and she gets good marks). Also, amongst those who bear witness to their experience, the majority are adults with a range of professional or technical skills, they are articulate about their situation and often bi-lingual. The greatest areas of shortfall, as revealed by these nine eco-docs, concern *networks*, *housing*, *health*, *income and work*, *social equity*, *political voice* and *peace and justice* – namely, the lack of infrastructures and support systems on the one hand and, on the other, the evacuation of any legal or democratic rights. In short: the unsafe, inhumane space of the unheard.

But even in this bleakest of chapters, and as indicated in the green biosphere above, there are glimmers of humanity's spirit shining through. The efforts of aid

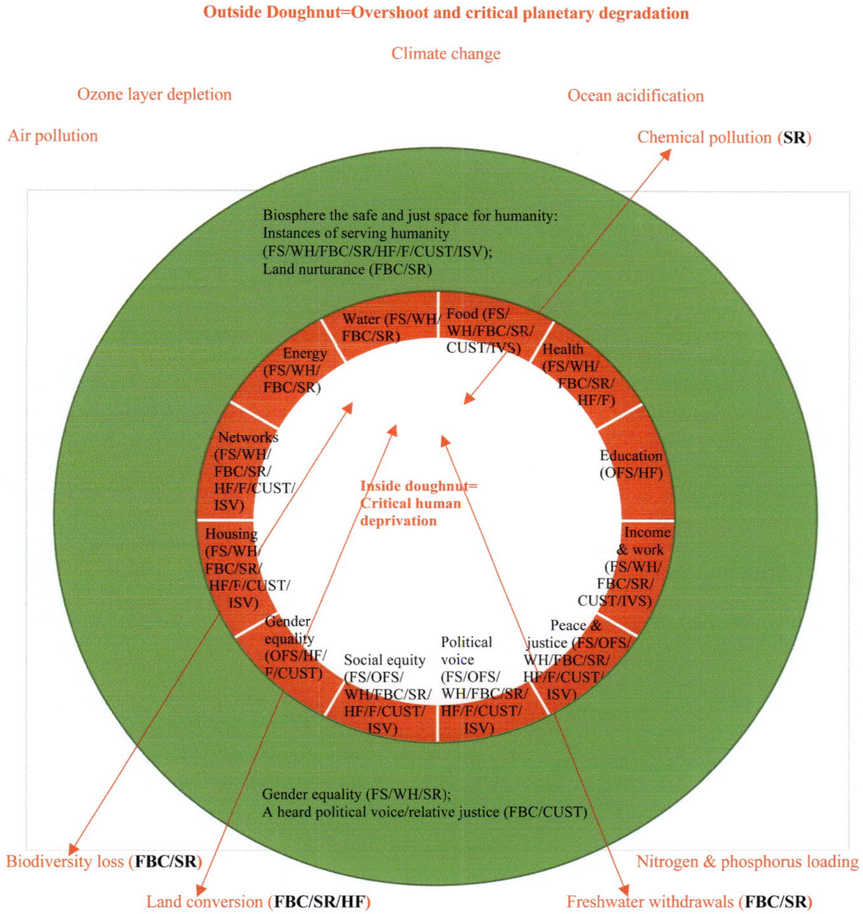

FIGURE 4.8 Mapping the nine eco-docs on disrupted ecosystems onto the Doughnut – *For Sama* (FS), *Of Fathers and Sons* (OFS), *White Helmets* (WH), *5 Broken Cameras* (FBC), *Samouni Road* (SR), *Human Flow* (HF), *Fuocoammare* (F), *Come un uomo sulla terra* (CUST) and *Il sangue verde* (ISV).

agencies, the compassion of doctors and hospitals, the lawyers putting themselves on the front line to plead the cause of those unjustly deprived of their rights, the extraordinary courage and resilience of those whose lives and livelihood have been assailed by the deregulated forces of those with the power to wreak havoc – this too is what we witness in these eco-docs. We observe also the gentle dignity with which these men, women and children who have been stripped bare, nonetheless, assert their humanity through their acts of kindness and humble heroism, their devotion to nurturing the Earth, their commitment to saving others, their peaceful demands for justice and equality. Last but not least, there is the dedication of the filmmakers who chose (or felt compelled by circumstances they could not ignore) to unlock the unseen truths, give space to the unheard voices and bring these narratives to our attention.

Notes

1 See: <https://www.un.org/sustainabledevelopment/blog/2019/05/nature-decline-unprecedented-report/> (accessed 15 February 2021).

2 This era is being seen as the sixth major extinction, but the only one that is directly caused by human behaviours. <https://en.wikipedia.org/wiki/Extinction_risk_from_climate_change> (accessed 15 February 2021).

3 From the 1880s to 1940, migration also went east to Asia from 'Old World' Europe (over 50 million to southeast Asia and another 50 million to central Asia). This was mostly driven by the global trade boom in commodities, however. Thus, in this instance, these are considered economic *migrants* who, for whatever reason, have not been qualified as refugees, although dire poverty was often the trigger. Further details can be found in: 'Two Centuries of International Migration', B. R. Chiswick & P. W. Miller (2013: 11–12) <http://ftp.iza.org/dp7866.pdf> (accessed 24 April 2021).

4 See: <https://www.unhcr.org/uk/statistics/country/556725e69/unhcr-global-trends-2014.html?query=refugee%20figures%202000-2014> and for a slightly different profile see: <https://openknowledge.worldbank.org/bitstream/handle/10986/25016/9781464809385.pdf> (both accessed 23 February 2021).

5 See: <https://www.unhcr.org/figures-at-a-glance.html> (accessed 15 February 2021).

6 Figures vary according to who you consult. Thus the International Organisation of Migration (IOM) places the figures between 20 and 37 million (see: <https://migrationdataportal.org/themes/irregular-migration>). But Wikipedia estimates the figure at around 60–70 million, citing India as in the tens of millions (20–30 million), the USA as 11 million and Russia as 10 million, and a further 7 countries absorbing some 11 million undocumented *migrants* (see: <https://en.wikipedia.org/wiki/Illegal_immigration>) (both accessed 20 March 2021).

7 See: <https://www.unhcr.org/news/stories/2001/6/3b4c06578/frequently-asked-questions-1951-refugee-convention.html> (accessed 22 February 2021).

8 See: https://www.unhcr.org/publications/legal/419c778d4/refugee-protection-international-law-article-31-1951-convention-relating.html (accessed 22 February 2021).

9 9 Just before this book goes to press the number of forcibly displaced people has risen to 82.4 million (with 26.4 million refugees). See: < https://www.unhcr.org/figures-at-a-glance.html> (accessed 17 August 2021).

10 Recently the mood in Uganda has changed drastically. Violence has broken out in response to protests against the current government and President Yoweri Museveni who have failed to deliver economically to the Ugandan people. Museveni came to power in 1986 and was a committed reformer (after the abuses of Idi Amin and Milton Obote). Since 2018, however, he has headed an increasingly repressive regime which leaves him accused of the same abuses as his predecessors. Presently, the Ugandan government has suspended 208 refugee aid organisations. This means there are just 69 such organisations authorised to assist the needs of 1.4 million refugees. See: <https://theconversation.com/why-uganda-has-suspended-hundreds-of-refugee-aid-agencies-145708> (accessed 23 April 2021).

11 For the full article on the *Memorandum of Understanding* see: Anja Palm (2017) <https://eumigrationlawblog.eu/the-italy-libya-memorandum-of-understanding-the-baseline-of-a-policy-approach-aimed-at-closing-all-doors-to-europe/> (accessed 16 Feb. 2021).

12 The quite scandalous situation in Turkey is not all of its making, of course. The prime culprit is the EU. But for a good overview of the situation, I draw your attention to this very clear report: <https://www.amnesty.org.uk/files/webfm/Documents/issues/embargoed_turkey_briefing_2june2016.pdf?wf7BB.dGPs9iTb9roUaznbQzJP9ivvdR> (accessed 12 February 2021).

13 See: <https://www.un.org/en/genocideprevention/war-crimes.shtml> (accessed 8 March 2021).

14 Figures cited in Samir Seifan (2011: 28), *The Road to Economic Reform in Syria*: <https://core.ac.uk/download/pdf/229308296.pdf> (accessed1 March 2021). Seifan provides a very full and clear study of this Liberalisation Campaign.

15 See: <https://reliefweb.int/report/syrian-arab-republic/fast-facts-syria-crisis-march-2019> and <https://www.statista.com/statistics/740245/number-of-internally-displaced-persons-in-syria/> (both accessed 1 March 2021).

16 See: <https://www.sciencedirect.com/science/article/pii/S0962629816301822> (accessed 1 March 2021).

17 As Wikipedia notes: 'the most intense drought ever recorded in Syria, which lasted from 2006 to 2011, resulted in widespread crop failure, an increase in food prices and a mass migration of farming families to urban centres. This migration strained infrastructures already burdened by the influx of some 1.5 million refugees from the Iraq war. The drought has been linked to anthropogenic global warming. Adequate water supply continues to be an issue in the ongoing civil war and it is frequently the target of military action'. <https://en.wikipedia.org/wiki/Syrian_civil_war> (accessed 1 March 2021).

18 *For Sama* lost out to *The Cave* (Feras Fayyad, 2019), a Danish-Syrian documentary about a doctor's courageous decision to continue operating in a makeshift hospital in Aleppo (what he called 'the cave'). The documentary stands as a record of the injustices committed against innocent victims of the Syrian Civil War.

19 At the time of the film's making there were about 111 local civil defence centres across 8 provinces; by 2017 this had grown to 179 centres. See: <https://en.wikipedia.org/wiki/White_Helmets_(Syrian_civil_war)> (accessed 5 March 2021).

20 By 2018 these numbers had grown to 114,000 rescued and 252 lost volunteers.

21 For a full review of Assad's use of chemical weapons since 2012 see: <https://www.armscontrol.org/factsheets/Timeline-of-Syrian-Chemical-Weapons-Activity> (accessed 5 March 2021).

22 For more information and to see the lengths to which Assad's regime will go in fuelling fake news against the White Helmets, consult the following webpages: <https://www.theguardian.com/world/2017/dec/18/syria-white-helmets-conspiracy-theories>;<https://en.wikipedia.org/wiki/White_Helmets_(Syrian_civil_war)>; <https://www.whitehelmets.org/en/>; and on the hounding of James Le Mesurier, the co-founder of the White Helmets, see: <https://www.bbc.co.uk/programmes/articles/8h3FB5DzDRN4pjZSCZX74K/why-no-one-could-save-the-man-who-co-founded-the-white-helmets> (all accessed 5 March 2021).

23 For a very full study see: <https://reliefweb.int/sites/reliefweb.int/files/resources/2019_Syr_HNO_Full.pdf> (the figures quoted for IDPs are on page 9) (accessed 5 March 2021).

24 See: <https://www.unhcr.org/sy/internally-displaced-people> (accessed 20 February 2021).

25 The following websites supply all of this information: <https://unesdoc.unesco.org/ark:/48223/pf0000371530>; <https://reliefweb.int/report/syrian-arab-republic/northwest-syria-newly-displaced-idps-assessment-december-2020> and <https://reliefweb.int/sites/reliefweb.int/files/resources/idp_map_23-02-2020.pdf> (all accessed 5 March 2021).

26 Source <https://reliefweb.int/sites/reliefweb.int/files/resources/E0507969E24DD371C1256F2D004800D5-israelcheckpoint_gaza.pdf> (accessed 9 March 2021).

27 Source: <https://en.wikipedia.org/wiki/Israelioccupied_territories#/media/File:Occupied_Palestinian_Territories.jpg> (accessed 9 March 2021).
28 Source: https://www.un.org/unispal/permanently-staffed-israeli-checkpoints-and-crossings-in-the-west-bank-ocha-map-as-of-may-2010/ (accessed 24 February 2021).
29 For full details see: <https://www.britannica.com/place/Palestine/International-recognition> (accessed 22 February 2021). For a recent short film on the effect of these internal checkpoints on Palestinians could I point you to the BAFTA short documentary-winning (2021) and Oscar-nominated *The Present*, Farah Nabulsi (2021).
30 See: <https://imeu.org/article/fact-check-msnbcs-palestinian-loss-of-land-map> and <https://www.bbc.co.uk/news/world-middle-east-29362505> (both accessed 22 February 2021). For the illegality of the Israeli West Bank Wall, see: <https://www.amnesty.org/en/latest/campaigns/2017/06/israel-occupation-50-years-of-dispossession/> (accessed 11 March 2021).
31 Source: <https://www.amnesty.org/download/Documents/48000/mde150152009en.pdf (accessed 9 March 2009)
32 Source: <https://www.amnesty.org/download/Documents/48000/mde150152009en.pdf (accessed 9 March 2009).
33 Beginning in the early 1970s, Israel took control of hundreds of thousands of dunams (acres) throughout the West Bank by declaring them state land. According to the applicable legislation, such a declaration is legal, inter alia, if it is proven that the land was not worked for at least three consecutive years. See: <https://www.btselem.org/settlements/20051228_modiin_ilit> (accessed 9 March 2021).
34 For a broad sweep of this contentious issue see: <https://en.wikipedia.org/wiki/Israeli_settlement> (accessed 9 March 2021). The International Court of Justice and the UN regard Israel as an occupying power, and deem it an affront to international law. See: <https://en.wikipedia.org/wiki/Israeli-occupied_territories#cite_note-mfafaq-11> (accessed 9 March 2021).
35 See: <https://unctad.org/press-material/despite-growth-long-term-prospects-palestinian-economy-worsened-report-says> (accessed 15 March 2021).
36 According to the UN, 56% of the Palestinian population is unemployed and 33% live below the poverty line. In Gaza the figures are the most extreme with 53% living below the poverty line. See: <https://reliefweb.int/sites/reliefweb.int/files/resources/Economic%20and%20social%20repercussions%20of%20the%20Israeli%20occupation%20on%20the%20living%20conditions%20of%20the%20Palestinian%20people%20in%20the%20OPT%2C%20including%20East%20Jerusalem%2C%20and%20of%20the%20Arab%20population%20in%20the%20occupied%20Syrian%20Golan.pdf> (accessed 9 March 2021).
37 Emad Burnat was later helped to edit the footage by Israeli filmmaker and supporter of the Bil'in cause, Guy Davidi (who, as a marker of solidarity between Palestinian and Israeli, is co-credited with the film).
38 See: <www.rv-press.com> (accessed 9 March 2021).
39 See: <https://www.britannica.com/place/Palestine/International-recognition> (accessed 9 March 2021).
40 For a full report, see: <https://www.amnesty.org/download/Documents/48000/mde150152009en.pdf> (accessed 12 March 2021).
41 See: <https://en.wikipedia.org/wiki/Gaza_War_(2008–2009)> (accessed 9 March 2021).
42 Source: <www.rv-press.com> (accessed 9 March 2021). In this regard, a significant moment towards the end of the film reveals a humane decision by one Israeli fighter pilot who refused, when ordered, to fire on civilians. Savona sums this up as follows: 'During the 30 seconds when the destinies of those people were decided, there was somebody who had to say yes or no. He said no and refused to obey orders, which shows that you always have a choice'. Interview: <https://www.haaretz.com/

middle-east-news/palestinians/MAGAZINE-an-animated-look-at-a-gaza-tragedy-1.6339853> (accessed 11 February 2021).

43 All quotations sourced from: <www.rv-press.com> (accessed 9 March 2021).

44 Image accessed: <https://www.nonfictionfilm.com/uploads/4/2/1/2/42121865/screenshot-2018-05-21-00-40-07_orig.png> (accessed 12 February 2021).

45 Sourced from *Samouni Road* Press Kit: <www.rv-press.com> (accessed 9 March 2021).

46 See: <https://www2.ohchr.org/english/bodies/hrcouncil/docs/12session/A-HRC-12-48.pdf> page 408. But see also pages 159–68 which deal with the Samouni case in detail and find no justification for the violence used by the Israeli military (accessed 18 March 2021).

47 Ibid., page 182.

48 See: <https://www.bbc.co.uk/news/world-middle-east-17602425> (accessed 18 March 2021).

49 For further details see: <https://www.amnesty.org.uk/truth-about-refugees#:~:text=Neither%20the%201951%20Refugee%20Convention,country%20in%20which%20they%20arrive> (accessed 27 April 2021).

50 See: <https://data.worldbank.org/indicator/SM.POP.REFG> (accessed 22 March 2021).

51 The UN's weakness on a number of issues is raised several times in this chapter and in Chapters 1 and 3. Former Special Rapporteur to the UN Human Rights Council, Agnès Callamard, puts it most clearly when she attributes this passivity to 'a mindset that is concerned about bothering the powerful'. See: <https://www.theguardian.com/law/2021/mar/23/agnes-callamard-human-rights-un-europe-amnesty-international> (accessed 23 March 2021).

52 See: <https://www.un.org/unispal/history> (accessed 12 March 2021).

53 For the full article see: <https://www.tni.org/en/publication/building-walls>; for the map see: <https://www.tni.org/files/building_walls_infographic_-_english.jpg> (both accessed 4 April 2021).

54 EU member states view the rescue activities of NGO rescue boats with hostility, and since 2016 have been pursuing a strategy of criminal and administrative proceedings against search and rescue (SAR) vessels. Despite this, the rescues carry on as best they can. More recently, in September 2020 the Banksy-funded rescue boat, Louise Michel, rescued 150 people. In November 2020 further grim figures emerged, 110 deaths at sea, a number lessened by the humanitarian intervention of the Open Arms ship. As of 2021, the previously detained SAR vessel, the German flag Sea-Watch 3 and her sister vessel 4 are amongst the ten actively deployed SAR vessels (a further five are being held in ports pending legal proceedings). See: <https://fra.europa.eu/en/publication/2020/2020-update-ngos-sar-activities> (accessed 28 April 2021).

55 See: < https://www.unhcr.org/africa.html> (accessed 22 March 2021).

56 For the Syrian numbers in Turkey, see: <https://www.unhcr.org/figures-at-a-glance.html>; for Lebanon, see: <https://www.unhcr.org/lb/wp-content/uploads/sites/16/2020/02/UNHCR-Lebanon-Operational-Fact-sheet-January-2020.pdf>; for Jordan see: <https://data2.unhcr.org/en/situations/syria/location/36>; for Europe, see: <https://www.pbs.org/wgbh/frontline/article/numbers-syrian-refugees-around-world/> (all accessed 22 March 2021).

57 See: <https://www.unhcr.org/ir/refugees-in-iran/> (accessed 22 March 2021).

58 See: <https://en.wikipedia.org/wiki/Immigration_to_the_United_States> (accessed 22 March 2021).

59 See: <https://en.wikipedia.org/wiki/Illegal_immigration#/media/File:Non-EU_citizens_found_to_be_illegally_present_in_the_EU-28_and_EFTA,_Eurostat_2015.png> (accessed 4 April 2021).

60 See: Ai Weiwei interview: <https://www.theguardian.com/commentisfree/2018/feb/02/refugee-crisis-human-flow-ai-weiwei-china> (accessed 17 February 2021).

61 A similar situation exists for refugees leaving Greece and taking the Balkan route towards the Croatia–Bosnia border in an endeavour to then cross either the Slovenian wall or the Serbian–Hungarian wall. Croatia becomes the border (albeit porous) that cannot be got through – the point of *refoulement*, therefore, exploited by the EU, as indeed are other borders on the Balkan route. As Lorenzo Tondo reports in *The Guardian*, 'many cases of *migrants* who reach Croatia are sent back to Bosnia by the authorities there. The same happens in Slovenia and Italy, where, last week, the court of Rome declared more than 700 pushbacks perpetrated by Italian police in Slovenia illegal. "Pushbacks are illegal, whether they are violent or not, it doesn't matter," says Nicola Bay (the Danish Refugee Council's Bosnia director). "They fundamentally undermine the right to international protection. Croatian pushbacks are a consequence of EU policy – aimed at transferring the responsibility for protecting people outside of the EU"' (parenthesis and italics mine). See: <https://www.theguardian.com/global-development/2021/feb/01/balkan-route-migrants-croatia-bosnia-border> (accessed 1 February 2021).

62 Bangladesh is not a signatory to the UN Refugee Convention of 1951 or 1967, and will no longer tolerate the Rohingya presence (1.1 million), so has begun the process of *refoulement*.

63 See: Ai Weiwei interview: <https://www.dw.com/en/ai-weiweis-human-flow-and-11-other-memorable-films-on-refugees/a-41392629> (accessed 17 February 2021).

64 See: Ai Weiwei interview: <https://nonfics.com/ai-weiwei-human-flow-interview-55477e19e525/> (accessed 17 February 2021, my parenthesis).

65 See: Ai Weiwei interview: <https://theculturetrip.com/europe/articles/ai-weiwei-tells-us-about-his-powerful-documentary-film-human-flow/> (accessed 17 February 2021).

66 I have written at greater length on *Fuocoammare*, a small part of which is reproduced here (Hayward, 2021).

67 Although Malta is considerably closer to Zuwara, very few immigrants make attempts to land there, well aware of the hostile reception they'll receive. Some immigrants seek to travel from Benghazi to Europe. The distance to Lampedusa is 488 miles (small wonder so many perish in the useless inflatable dinghies they are often forced to travel upon). Crete is closer, but the story of drowning is much the same.

68 Italy operated the *Mare Nostrum*, whose reach went as far as the Libyan coast. The SAR operations it conducted saved over 150,000 lives. Although the European Commission provided financial support, the expense was ultimately shouldered by Italy. Italy requested support from the other EU member states, but was refused. So *Mare Nostrum* was rescinded. When Frontex became the sole operator, their fleet was of far smaller range, remaining closer to the Italian coast. Its focus was border protection rather than search and rescue.

69 During his various periods in office, Prime Minister Berlusconi signed a number of deals with the then-Libyan leader Colonel Gaddafi (the latest one in the context of *Fuocoammare* being 2008), by which *migrants* were detained in Libya in atrocious conditions only to then be sent back to their country of origin, or forced into prostitution, or killed 'trying to escape'.

70 Once afloat on these vessels, some of the women die having given birth to their child, also dead and still attached to its mother by its umbilical cord – and so it goes on.

71 In the Mediterranean area alone it is estimated that people smugglers pocket around €100 million yearly. See: <https://frontex.europa.eu/media-centre/news/news-release/after-the-money-prices-for-people-smuggling-on-central-and-western-mediterranean-routes-EHDfJg> (accessed 25 March 2021).

72 This lucrative 'business' is of course still ongoing. In the first eight months of 2020 alone, according to the International Organisation for Migration, '76,000 *migrants* have been intercepted so far this year and returned to Libya'. In the same period, 'more than 500 refugees and *migrants* are known to have died in the Mediterranean ... and more than 19,500 *migrants* have survived ... and reached Italy or Malta' (italics

mine). See: <https://www.theguardian.com/world/2020/aug/27/banksy-funds-refugee-rescue-boat-operating-in-mediterranean> (accessed 27 August 2020).

73 It might interest you to know that two successive mayors of Rosarno, Carlo Martelli and Giuseppe Ìda, have been arrested for illegal practices involved with the mafia. For Martelli, no case was found to answer. Ìda is currently under house arrest.

74 See: Segre's blog: <http://ilsangueverde.blogspot.com/p/english-info.html> (accessed 30 March 2021).

75 Cited in <https://www.cram.com/essay/The-Rosaro-Riots-In-Rosarno-Italy/FKKBZZ5KGYKW> (accessed 30 March 2021).

CONCLUSION

This is a conclusion only insofar as it is the end of the book. In the eco-docs discussed in the preceding chapters, everything that needed to be said has been raised with regard to our relentless drive, over the last 200 years, for progress and capital gain at all costs to the Earth, at the expense of an encompassing commitment to the well-being for all species inhabiting this world and, too, the planet itself. We have known that this was a fateful pathway to choose as early as the 1960s, when Rachel Carson, in *The Silent Spring*, declared:

> We stand now where two roads diverge. But unlike the roads in Robert Frost's familiar poem, they are not equally fair. The road we have long been traveling is deceptively easy, a smooth superhighway on which we progress with great speed, but at its end lies disaster. The other fork of the road – the one less traveled by – offers our last, our only chance to reach a destination that assures the preservation of the earth.[1]

Yet, as we have blindly trundled down the path of our own destruction, we have always known – certainly over the last 60 years – exactly how we could heal the Earth and thereby also heal ourselves. Scientists, humanists, ecologists and more recently economists have all made the same point – capitalism, neoliberal globalisation, the GDP growth model, monopolistic practices, etc., do not work. Do not work, that is, if we wish to survive. When Carson poses the question, 'how could intelligent beings seek to control a few unwanted species by a method that contaminated the entire environment and brought the threat of disease and death even to their own kind', we have to concur with both her query and her response which is to ask 'whether any civilisation can wage relentless war on life without destroying itself, and without losing the right to be called civilised'.[2]

DOI: 10.4324/9781003130246-102

Franny Armstrong's *Age of Stupid* (2009) pursues this questioning, only to conclude that 'we could have saved ourselves, but we didn't'. In her futuristic documentary, set in 2055, about our failure to preserve the environment, Armstrong draws upon mainstream scientific projections to show – via her on-screen narrator, and sole human survivor, Pete Postlethwaite – how the Anthropocene era *is* the Age of Stupid. How could we claim not to know that our disrespect for the Earth, our calamitous pursuit of 'taming nature' and our complacent self-assurance that what is out there is ours to take, would eventually bring us to the brink of finitude? The planet is one, and can nurture us only if we accept Earth's carrying-capacity conditions. There are not 'five Earths' to satisfy our insatiable desires. This Anthropocene era is the age in which, particularly in wealthy nations, our refusal to acknowledge the harm of our deregulated madness (in the form of addiction to fossil-fuel energy, over-consumption, wealth accumulation, exploitative practices, etc.) has caused temperatures to rise almost to tipping point, so that, come 2055, it is highly likely it will be too late to turn the tide of the consequences of humankind's massive contribution to global warming. Given the Paris Agreement of 2015, which is a legally binding international treaty on climate change, and which commits the signatories to limit global warming to well below 2 degrees Celsius compared to pre-industrial levels, by a reduction of carbon emissions of 70% as of 2030, how is it possible to square *that* commitment with the fact that, as of 2021, carbon emissions are set to jump this year by the second-highest rise in history as global economies pour stimulus cash into fossil fuels (as part of a recovery plan for this mega-rich industry to recover from the impact of the COVID-19 recession)?[3] All this tells us is that oil and coal lobbies are still pulling on the ears of governments and investors, and deliberately (contemptuously) ignoring any internationally agreed targets of carbon reduction.

As confirmation of the incontrovertible truth of the damage human progress has wrought, Luc Jacquet's *La Glace et le ciel/Ice and the Sky* (2015) goes back in time to provide, through the work of the glaciologist Claude Lorius, the evidence that the Earth's natural cycles of glacial and inert-glacial eras have been disturbed by anthropogenic global warming. In 1957, as part of the International Geophysical Year, Lorius went on an expedition to the Antarctic about which very little was then known. This was the beginning of some 20 visits Lorius was to make over a period of more than 30 years, during which time he made scientific discoveries about the cyclical nature of the glacial and interglacial epochs and provided evidence that the climate and concentrations of greenhouse emissions have always been inextricably linked. This work consisted of drilling out an ice core from the Antarctic ice cap and retrieving cylinders of ice for study. These samples of compressed snow, which have accumulated over tens of thousands of years, allowed Lorius to reconstruct the specific climates corresponding to those years. His breakthrough came when he noticed that as the ice melted so it released small bubbles of air. As he explains 'ice, when it melts, lets out the air with which it landed; so when it releases that air, it's the same as the snow particle'. These bubbles constitute 'atmospheric fossils', he informs us, which was an amazing discovery. 'The isotopic thermometer was born', he exclaims, and 'tells us precisely when that particular particle of snow fell'. Thus, by analysing the bubbles taken from the whole thickness of the ice (brought up in the cylinders from drilling), Lorius was able 'to reveal the history of the climate since the dawn of time'. Over subsequent visits, right through to 1991, Lorius went ever deeper — assisted of course by the work of many other glaciological groups, including the former USSR's Antarctic Vostock Centre whose advanced technology facilitated his last two drilling analyses (1988 and 1991), allowing Lorius to confirm the findings of his earlier drillings (which were then based on a sample of a mere 40,000 years). These later drillings provided, respectively, 420,000 and 800,000 years of ice and proved that, in terms of the Earth's eternal course around the sun, there is a 100,000-year cycle of cold and warm phases (an 80,000-year glacial followed by a 20,000-year interglacial period). Lorius studied these samples, first in 1988, which evidenced the regularity of the 100,000-year cycle of four climactic cycles within the 420,000-year-old sample. This was followed by the eight cycles in the later 1991 drilled sample (of 800,000 years). Lorius was able to prove, graphically, that the temperature varies naturally on Earth of up to 5 degrees Celsius between its cold (glacial) and warm (interglacial) periods. In other words, when CO_2 levels are low the planet cools down (ice age); when they are high the planet heats up (interglacial age). He was also able to show that sea levels varied by up to 120 metres depending on whether the water was frozen in ice caps or free in liquid form.

However, what Lorius also discovered, already in the first 30 years of his analysis, was 'how man through burning coal and wood was managing to change the Earth's climate'. He further discovered, as he tells us,

> that during the nuclear explosion experiments a certain number of radioactive elements released into the atmosphere returned to the earth around the explosion; others went up into the upper atmosphere and circulated around our globe and these are the radioactive elements in particular we have found in the layers of snow at the polar regions.

He was able to pinpoint the precise timing of the release of the nuclear bombs: 'in this white wilderness, never colonised by man, lies humanity's signature, Nagasaki, Hiroshima, every nuclear explosion since the discovery of nuclear weapons'.[4]

From his early optimistic investigations of the late 1950s, this eminent glaciologist came to the devastating realisation of 'how finite the planet is, given our behaviour and destructive omnipresence'. As his scientific study of the top layers of the ice core makes clear

> the message is irrefutable; the truth remains that over the past 200 years, the CO_2 produced by man is behind our unprecedented rise in temperatures

> on Earth. We are altering our planet's climate at an unprecedented rate never before in history.

In short we are transgressing Earth's natural temperature variant, thereby heating the planet up at an alarming rate – which is why we need to move swiftly to rebalance the planet's equilibrium and approximate the natural order of the Earth's climactic cycles.

At the time of the making of this film, Lorius was 83. Remarkably, he still has hope for humanity. And so I will end this book with his pronouncement at the end of *La Glace et le ciel*:

> The human is never more sublime than when faced with adversity, now you know the state we are in, what are you going to do?

Notes

1 Citation from: <https://www.goodreads.com/author/quotes/15332.Rachel_Carson> (accessed 4 May 2021).
2 Ibid.
3 See: <https://www.theguardian.com/environment/2021/apr/20/carbon-emissions-to-soar-in-2021-by-second-highest-rate-in-history> (accessed 20 April 2021). The melting of glaciers has doubled since the new millennium; with it, we can anticipate huge flooding problems in the near future, beyond the flooding already occurring; as is the already diagnosed tilting of the polar axis. See: <https://www.theguardian.com/environment/2021/apr/28/speed-at-which-worlds-glaciers-are-melting-has-doubled-in-20-years> and <https://www.theguardian.com/environment/2021/apr/28/what-we-lose-when-we-lose-our-glaciers> (both accessed 28 April 2021).
4 According to a 2002 study, the Arctic environment is also affected by radioactive pollution which means that indigenous people who live mainly on traditional food products (from the local sea-waters) receive the highest doses of radioactive contamination. See: <https://inis.iaea.org/collection/NCLCollectionStore/_Public/33/071/33071236.pdf> (accessed 9 May 2021).

BIBLIOGRAPHY

Acemoglu, Daron & Robinson, James A. (2013) *Why Nations Fail: The Origins of Power, Prosperity and Poverty*, London, Profile Books.

Angelone, Anita (2011) 'Talking Trash: Documentaries and Italy's "garbage emergency"', *Studies in Documentary Film*, 5:2/3, 145–56.

Bozak, Nadia (2011) *The Cinematic Footprint: Lights, Camera, Natural Resource*, New Brunswick & London, Rutgers University Press.

Corrado, Alessandra (2011) '*Clandestini* in the Orange Towns: Migrations and Racisms in Calabria's Agriculture' *Race/Ethnicity: Multidisciplinary Global Contexts*, 4:2, 191–201. https://doi.org/10.2979/racethmulglocon.4.2.191 (Accessed 30 March 2021).

Demos, T. J. (2017) *Against the Anthropocene: Visual Culture and Environment Today*, Berlin, Sternberg Press.

Duvall, John A. (2017) *The Environmental Documentary: Cinema Activism in the 21st Century*, London, Bloomsbury.

Ferris, Elizabeth & Kirisci, K. (2016) *The Consequences of Chaos: Syria's Humanitarian Crisis and the Failure to Protect*, Washington, Brookings Institute Press Marshall Papers Series.

Gagnier, Regenia (2000) *The Insatiability of Human Wants*, Chicago, University of Chicago Press.

Hayward, Susan (2020) *Film Ecology: Defending the Biosphere – Doughnut Economics and Film Theory and Practice*, London & New York, Routledge Focus Series.

Hayward, Susan (forthcoming) 'European-documentaries as Performative Exchange', In: Gábor Gergely & Susan Hayward (eds), *Companion to European Cinema*, London, Routledge.

Hjort, Mette (ed.) (2012) *Film and Risk*, Detroit, Wayne University Press.

Hughes, Helen (2014) *Green Documentary: Environmental Documentary in the Twenty-First Century*, Bristol & Chicago, Intellect Press.

Kimmerer, Robin Wall (2020) *Braiding Sweetgrass*, London, Penguin Books (first printed 2013, by Milkweed Editions, Minneapolis).

Livingstone-Smith, David (2011) *Less than Human: Why We Demean, Enslave and Exterminate Others*, New York, St Martins Press.

Millar, Kathleen M. (2018) *Reclaiming the Discarded: Life and Labor on Rio's Garbage Dump*, Durham & London, Duke University Press.

O'Brien, Adam (2018) *Film and the Natural Environment: Elements and Atmosphere*, London & New York, Wallflower Press.

Paz, Moria (2017) 'Law of the Walls', *European Journal of International Law*, 28:2, 601–24.

Raworth, Kate (2017) *Doughnut Economics: Seven Ways to Think Like a 21st- Century Economist*, London, Random House Business Books.

Rust, Stephen, Monani, Salma & Cubitt, Sean (2013) *Ecocinema: Theory and Practice*, London & New York, Routledge.

APPENDIX A

List of 57 European ecology-documentaries 2000–19

2000	*Les Glaneurs et la glaneuse/The Gleaners and I* (Fr) Of note: world-wide festival honours including Sélection *Hors competition* Cannes 2000; eight critics' awards including Prix Méliès 2000; $159,165 box office	The wide variety of gleaning practices, from agricultural leftovers to urban plastic waste, and the uses they can be put to	Agnès Varda
2004	*Darwin's Nightmare* (Austrian/Fr/Bel) Of note: César best documentary 2006; numerous festival awards; Oscar nomination for best documentary 2005; $3 million box office	Environmental and social effects of fishing industry around Lake Victoria, Tanzania	Hubert Sauper
2005	*Before Vanishing* (Fr/Syria)	Urbanisation/pollution along Barada river Syria	Joude Gorani
	McLibel (UK) Of note: Grierson Award for Best Documentary 2005; British Independent Films Award 2005	McDonald's big corporate practices taken to task (jobs, food, environmental practices)	Franny Armstrong and Ken Loach

(continued)

2006	*A Crude Awakening: The Oil Crash* (Switzerland) Of note: international festivals including Edinburgh Film Festival; several minor festival awards	Impact of oil dependency (world-wide)	Basil Gelpke and Ray McCormack
2007	*Earth* (Fr/UK/Ger/US) Of note: $109 million box office; minor awards	Climate change impact on species survival (world-wide)	Alastair Fothergill and Mark Linfield
	Biùtiful cauntri/Beautiful Country (It) Of note: Italian Golden Globe best documentary award 2008; and two other festival awards; $141,693 box office	Pollution/illegal toxic waste disposal in Italy	Esmeralda Calabria and Andrea D'Ambrosio
	Clouds of Smoke (Alb/UK)	Global warming impact in Albania	Fatmir Terziu
2008	*Come un uomo sulla terra/Like a Man on Earth* (It) Of note: Academy of Italian Cinema, David di Donatello award for best documentary 2009	Ethiopian refugees living in Rome; brutal controls of Libya and Italy controlling flow of African migrants to Europe	Andrea Segre, Dagmawi Yimer, Ricardo Biadene
	Forgotten District (Fr/UK/Ger)	Ecotourism/saving rainforest in Belize	Oliver Dickinson
	La Bambina deve prendere aria/Baby Needs Some Fresh Air (It)	Air pollution in Italy	Barbara Rossi-Prudente
	Les Enfants nous accuseront/That Should Not Be: Our Children Will Accuse Us (Fr)	Agriculture: impact of agrochemicals on food production	Jean-Paul Jaud
	Le Monde selon Monsanto/The World According to Monsanto (Fr) Of note: Cannes Trophée des sciences du danger 2008; Rachel Carson Prize; minor festivals and awards	Use and promotion of genetically modified seeds (world-wide)	Marie-Monique Robin
	La Vie moderne (Fr) Of note: Prix Louis-Leduc 2008; nominated for César 2009; $2.4 million box office	Testimonies of mountain-living peasants about their rural existence and farming life	Raymond Depardon

2009	*Age of Stupid* (UK) Of note: several minor festival awards; $346,176 box office	Climate change (world-wide)	Franny Armstrong
	Dreamland (Iceland) Of note: $123,238 box office	Environmental preservation: effect of dams and hydropower; links of 'green power' with giant corporations	Andri Snaer Magnason
	Forests of the 21st Century (UK)	Forests: reforestation, carbon emissions	Forestry Commission of Great Britain
	Home (Fr) Of note: César best documentary 2010; two other international awards; $1.9 million box office	Impact of human civilization on the continents of the world (world-wide)	Yann Arthus-Bertrand
	The Idiot Cycle (Fr/Can) Of note: Romanian Film Festival award 2009	Chemical industry and GMOs and their impact on humans (cancer)	Emmanuelle Schick-Garcia
	Una montagna di balle/A Mountain of Lies (It)	Illegal disposal of toxic waste in Italy	Nicola Angrisano
	Vanishing of the Bees (UK) Of note: two minor festival awards	Honey bees: colony collapse disorder	Maryam Henein, George Langworthy and James Erskine
2010	*The 4th Revolution* (Ger) Of note: minor festivals; $219,694 box office	Renewable energy	Carl-A 1
	Il sangue verde/Green Blood (It/Fr) Of note: Venice Film Festival, Venice Days Selection 2010; minor festival awards	Migrant labour revolts	Andrea Segre
	Into Eternity (Den) Of note: minor festival award	The construction of Onkalo nuclear waste depository in Finland; investigating the concept of long-term (eternal) storage of radioactive waste	Michael Madsen
	No Pressure (UK)	Climate change	Dougal Wilson

	Solutions locales pour un désordre local/Think Global, Act Local (Fr) Of note: $107,831 box office	The impact of big agribusiness versus the organic approach to food production (world-wide)	Coline Serreau
	Waste Land (UK/Brazil) Of note: 50 awards (festivals and associations); Oscar nomination for best documentary; $291,307 box office	Recycling landfill into contemporary art	Lucy Walker
2011	*Five Broken Cameras* (Fr/Israel/Netherlands/UK/ Finland/ Switzerland/US/ Can/Rep of Korea) Of note: Oscar nomination 2012; three major festival awards and five minor festival awards; $146,475 box office	Bearing witness to the non-violent Palestinian resistance in Bil'in, a West Bank village gradually encroached upon by Israeli settlements	Emad Burnat and Guy Davidi
	Just Do It: A Tale of Modern Day Outlaws (UK)	Non-violent civil disobedience by campaigners against global warming	Emily James
2012	*More than Honey* (Switzerland) Of note: minor festival awards; $1.6 million box office	Honey bee colonies in various countries (world-wide)	Marcus Imhoof
	Polluting Paradise (Ger) Of note: Special Screenings Selection Cannes 2012	Waste management in Turkey	Fatih Akin
	Solar Mamas (Den/UK/US) Of note: Peabody Award 2012; several minor awards; minor festivals	Women from Jordan, Kenya, Columbia and Burkina Faso travel to India to learn the basics of solar engineering, returning to their countries bringing the technology to their villages	Jehane Noujaim and Mona Eldaief
	Stealing Africa (Den) Of note: minor festivals and awards	Multinational British commodity trading company Glencore's (predominantly Swiss-based) exploitation of copper mines in Zambia at the expense of Zambians	Christoffer Guldbrandsen

	Trashed (UK) Of note: won several awards – Earth Grand Prix award at Tokyo International Film Festival 2012; premiered at Cannes 2012, nominated for the Caméra d'Or	Global impact of humanity's wasteful consumerism and pollution	Candida Brady
	Cultures en transition/Voices of Transition (Fr/Ger) Of note: shown at most of the environmental film festivals	Agriculture, agroecology and solutions to energy and resource scarcity	Nils Aguilar
2013	*Au nom de la terre/For the Sake of the Earth* (Fr) Of note: minor festivals	Pioneering work of Pierre Rabhi, an agri-ecologist for 60 years, offering a new model of society	Marie-Dominique Dhelsing
	Journey to the Safest Place on Earth (Switzerland) Of note: minor festivals	Radioactive waste	Edgar Hagen
2014	*Anaïs s'en va-t-en guerre/Anaïs Goes to War* (Fr) Of note: minor festivals	A young woman seeks an alternative life cultivating herbs and aromatic plants	Marion Gervais
	Broken Land (Switzerland) Of note: minor film festivals	Migrants at Mexico/USA wall	Stéphanie Barbey and Luc Peter
	Le Sel de la terre/The Salt of the Earth (Fr/Ger) Of note: César for best documentary 2015; Special Prize at Cannes Film Festival 2014; two festival awards; Oscar nominated; $5.4 million box office	Exploitation of environment and humans to maximise profit for the global economic market (life and work of Sebastiao Salgado) (world-wide)	Wim Wenders and Juliano Ribeiro Salgado
	Virunga (UK/Congo) Of note: Oscar nominated 2015; Peabody Award 2014; Emmy Award 2015; several minor festival awards	Abuse of wildlife conservation practices in Virunga National Park (Democratic Republic of Congo)	Orlando Von Einsiedel
2015	*The Day the Sun Fell* (Switzerland) Of note: 11 minor festival awards; selected for 19 minor festivals	Witness survivors of and ongoing impact of USA atomic bombing of Hiroshima in 1945	Aya Domenig

	Demain/Tomorrow (Fr) Of note: César best documentary 2016; $11 million box office	Sustainable communities (world-wide)	Cyril Dion and Mélanie Laurent
	La Glace et le ciel/Ice and the Sky (Fr) Of note: Closing Film of Cannes Film Festival 2015	Global warming (Antarctic) About the glaciologist Claude Lorius' findings	Luc Jacquet
	How to Change the World (UK/Can) Of note: World Cinema Documentary Special Jury Award at Sundance 2015; five other minor awards	Environmentalism: Greenpeace	Jerry Rothwell
2016	*Fuocoammare/Fire at Sea* (It) Of note: Golden Bear award at Berlin Film Festival 2016; Oscar nominated 2016; Master in Conversation at Venice Biennale 2016; $1.2 million box office	European migrant crisis	Gianfranco Rosi
	Humain/Human (Fr) Of note: Vancouver Film Festival award for most popular international documentary 2016; Venice Biennale 2016	Human testimonies from 60 countries (exile, displacement, etc.) (world-wide)	Yann Arthus-Bertrand
	White Helmets (UK) Of note: Oscar award for best short documentary 2017	Daily operations of the Syrian Civil Defence	Orlando Von Einsiedel
2017	*Human Flow* (Ger) Of note: Golden Lion Venice 2017; several human rights awards; $1.8 million box office	Global refugee crisis	Ai Weiwei
	Of Fathers and Sons (Ger/Syria) Of note: World Cinema Documentary Grand Jury Prize, Sundance Film Festival 2018; Oscar Nomination 2018	Jihadi recruitment and terrorist training in Syria	Talal Derki

	Round-up face à ses juges/ Round-up facing its judges Of note: Rachel Carson Prize (for film and book of the film)	Monsanto brought before the International Monsanto Tribunal	Marie-Monique Robin
	Sea Sorrow (UK) Of note: Special Screening Selection at Cannes Film Festival 2017	Child refugees in the European migrant crisis	Vanessa Redgrave
2018	*Après demain* (Fr)	Follow up on *Demain*, documenting successful and not so successful agricultural practices	Cyril Dion and Laure Noualhat
	Eldorado (Switzerland) Of note: Amnesty International Film Prize Berlin 2018; minor film festivals and awards	The European migrant crisis	Markus Imhoof
	Samouni Road (It/Fr) Of note: Cannes Best Documentary Award (*Oeil d'Or*) 2018; David di Donatello Award for Best Documentary 2018	Impact of Israeli bombarding (the Zeitoun Incident of December 2008–January 2009) on the Palestinian Gaza Strip	Stefano Savona
	Undamaged (Slovenia)	Measures being taken to protect wild rivers in the Balkan states	Miha Augustin, Rozle Bregar and Matik Oblak
2019	*For Sama* (UK) Of note: BAFTA for best documentary 2019; four British Independent Film Awards 2019; Oscar nominated 2020; $1.4 million box office	Syrian civil war	Waad Al-Kateab and Edward Watts

INDEX

Index of film titles

The Age of Stupid (2009) 3, 120, 126, 169, 177
Anaïs s'en va-t-en guerre/Anaïs Goes to War (2014) 3, 84, 85, 108, 179
Après demain/After tomorrow (2018) 3, 84, 85, 181
Atlantis (1991) 51n7
Au nom de la terre/For the Sake of the Earth (2013) 3, 4, 84, 85, 100–8, 179

La bambina deve prendere aria/Baby Needs Some Fresh Air (2008) 3, 60, 62, 63–71, 176
Before Vanishing (2005) 3, 60, 175
Biùtiful cauntri/Beautiful Country (2007) 3, 4, 60, 62, 63–71, 176
Broken Land (2014) 3, 120, 153, 179

The Cave (2019) 163n18
Clouds of Smoke (2007) 3, 60, 63, 71, 82n4, 176
Come un uomo sulla terra/Like a Man on Earth (2008) 3, 4, 120, 123, 124, 153, 155–7, 160–2, 176
A Crude Awakening: The Oil Crash (2006) 3, 15, 20, 25–6, 37–8, 176
Cultures en transition/ Voices of Transition (2012) 3, 84, 85, 100–3, 108–10, 112, 179

Darwin's Nightmare (2004) 3, 4, 85, 114n2, 175
The Day the Sun Fell (2015) 3, 5, 6, 15, 20, 21, 22–5, 37–8, 44, 181
Demain (2015) 3, 4, 84, 85, 100–3, 110–11, 180
Dreamland (2009) 3, 15, 17, 20, 27–9, 34, 36–8, 177

Earth (2007) 3, 4, 6, 85, 176
Eldorado (2018) 3, 5, 120, 153, 155, 181
Les Enfants nous accuseront/Our Children Will Accuse Us (2008) 3, 85, 176

5 Broken Cameras (2011) 3, 120, 133–41, 146, 148, 160–2, 178
Forests of the 21st Century (2009) 3, 84, 85, 177
Forgotten District (2008) 3, 84, 85, 176
For Sama (2019) 3, 4, 120, 126–8, 130, 132, 160–2, 163n18, 181
The 4th Revolution (2010) 3, 15, 40–5, 48–51, 56n, 56n70, 177
Fuocoammare/Fire at Sea (2016) 3, 4, 120, 123, 124, 153–6, 160–2, 166n66, 166n69, 167n73, 167n75, 180

La Glace et le ciel/Ice and the Sky (2015) 3, 120, 126, 169–71, 180
Les Glaneurs et la glaneuse/The Gleaners and I (2000) 3, 6, 15, 39–40, 175
Le Grand bleu (1988) 51n7

Hiroshima mon amour (1959) 53n23
Home (2009) 3, 4, 6, 15, 16–19, 51n3, 125, 177

How to Change the World (2015) 3, 4, 15, 180
Humain/Human (2016) 3, 6, 85, 180
Human Flow (2017) 3, 4, 10, 120, 123, 124, 147, 149, 151–3, 160–2, 166n63, 181

Idiot Cycle (2009) 3, 85, 94, 98, 116n37, 177
Into Eternity (2010) 3, 15, 20, 21–2, 24–5, 36–8, 53n22, 177

Journey to the Safest Place on Earth (2013) 3, 15, 179
Just Do It (2011) 3, 85, 178

Léon (1994) 51n7
Lords of Water (2020) 101, 117n48

McLibel (2005) 3, 85, 175
Le Monde selon Monsanto/The World According to Monsanto (2008) 3, 4, 6, 85, 88–94, 98– 100, 115n14, 176
Una montagna di balle/Mountain of Lies (2009) 3, 60, 62, 63–71, 80–1, 177
More than Honey (2012) 3, 85, 95–7, 98–100, 178

Nikita (1990) 51n7
No Pressure (2010) 3, 60, 82n3, 178

Of Fathers and Sons (2017) 3, 4, 120, 126–32, 160–2, 181

The Present (2021) 164n29
Polluting Paradise (2012) 3, 60, 63, 71, 82n5, 178

Round-up face à ses juges/Round-up Facing Its Judges (2017) 3, 85, 88–94, 98–100, 181

Samouni Road (2018) 3, 4, 120, 132, 133, 134, 137, 141–7, 160–2, 165n45, 181
Il sangue verde/Green Blood (2010) 3, 120, 124, 153, 155–7, 160–2, 167n75, 177
Sea Sorrow (2017) 3, 120, 124, 153, 181
Le Sel de la terre/The Salt of the Earth (2014) 3, 4, 85, 179
Solar Mamas (2012) 3, 15, 37, 40–50, 57n83–4, 178
Solutions locales pour un désordre global/ Think Global, Act Local (2010) 3, 84, 85, 100–7, 178
Stealing Africa (2012) 3, 15, 27, 29–32, 36–8, 55n49–50, 179

Trashed (2012) 3, 60, 62, 71, 81, 179

Undamaged (2019) 3, 5, 12n2, 15, 27, 29, 36–8, 54n41, 181

Vanishing of the Bees (2009) 3, 85, 95–100, 177
La Vie moderne/Modern Life (2008) 3, 84, 85, 176
Virunga (2014) 3, 15, 27, 29, 32–8, 55n54–7, 56n58–9, 56n65, 130, 180

Waste Land (2010) 3, 12n1, 40, 60, 62, 73–81, 83n23, 178
White Helmets (2016) 3, 4, 120, 126, 128, 130–2 160–2, 163n19, 163n22, 180

Zambia: Good Copper Bad Copper (2011) 54n45

Index of film directors

Aguilar, Nils 102, 108–10, 118n66, 179
Akin, Fatih 178
Al-Kateab, Waad & Edward Watts 126, 128, 181
Angrisano, Nicola 67, 177
Armstrong, Franny 169, 177
Armstrong, Franny & Ken Loach 175
Arthus-Bertrand, Yann 6, 16–17, 177, 180
Augustin, Miha, Rozle Bregar & Matik Oblak 29, 181

Barbey, Stéphanie & Luc Peter 153, 179
Brady, Candida 60, 81n2, 179
Burnat, Emad & Guy Davidi 138–41, 164n37, 178

Calabria, Esmeralda & Andrea D'Ambrosio 66, 176

Dhelsing, Marie-Dominique 107–8, 179
Depardon, Raymond 176
Derki, Talal 128–30, 181
Dickinson, Oliver 176
Dion, Cyril & Mélanie Laurent 109–11, 180
Dion, Cyril & Laure Noualhat 181
Domenig, Aya 22–3, 180

Fetchner, Carl-A 40–4, 177
Forestry Commission of Great Britain 177
Fothergill, Alastair & Mark Linfield 6, 176

Gelpke, Basil & Ray McCormack 25, 176
Gervais, Marion 179

Gorani, Joude 175
Guldbrandsen, Christoffer 30, 179

Hagen, Edgar 179
Henein, Maryam, George Langworthy & James Erskine 95, 177

Imhoof, Marcus 95, 153, 155, 178, 181

Jacquet, Luc 169, 180
James, Emily 178
Jaud, Jean-Paul 176

Madsen, Michael 21–2, 53n22, 177
Magnason, Andri Snaer 27, 177

Noujaim, Jehane & Mona Eldaief 40, 178

Redgrave, Vanessa 124, 153, 181
Robin, Marie-Monique 88–94, 113–14n1, 116n25, 176, 181
Rosi, Gianfranco 153, 180
Rossi-Prudente, Barbara 67–8, 176
Rothwell, Jerry 180

Sauper, Hubert 175
Savona, Stefano 141–6, 164n42, 181
Schick-Garcia, Emmanuelle 94, 116n36, 177
Segre, Andrea 155, 156–9, 167n74, 177
Segre, Andrea, Dagmawi Yimer, Ricardo Biadene 155, 156, 176
Serreau, Coline 104–6, 178

Terziu, Fatmir 63, 82n4, 176

Varda, Agnès 6, 39–40, 175
Von Einsiedel, Orlando 33–5, 56n59, 56n61–2 130–1, 180
Walker, Lucy 73, 74, 77, 178
Weiwei, Ai 10, 124, 147, 149, 151–3, 165n60, 166n63–5, 181
Wenders, Wim & Juliano Ribeiro Salgado 179
Wilson, Dougal 178

Handy index to Doughnut model and international bodies

Doughnut Economics/Doughnut model/ diagram 1, 2, 7–12, 16, 18, 36–8, 44, 48–51, 59, 71–3, 79, 80, 98–100, 111–13, 118n, 125–6, 132, 139–41, 146, 160–2, 173, 174

International bodies (including giant corporations, patenting practices, treaties)

ALCOA 27–9, 54n39
Amnesty International 3, 131, 164n30–2, 164n40, 165n49, 181
Astrazeneca 94, 116n36

BASF 94, 113–14n1
Bayer 89, 94, 95, 105, 113–14n1, 115n19

ChemChina 89, 97, 105

Dow Chemical 89, 90, 94, 95, 105, 113–14n1
DuPont 89, 94, 105, 113–14n1

EIB/European Investment Bank 31, 54n48, 55n49
European Commission/European Union (EU) 64, 68, 70, 82n6, 83n25, 89, 95, 101, 108, 115n11, 116n26, 123–4, 147, 149, 150, 151, 153–9, 163n12, 165n54, 166n61

FDA/Food and Drug Administration 92, 93, 95, 101
FTA/Free Trade Agreement 89

Geneva Convention 137, 142, 147, 160
Glencore 30, 31, 54n44, 55n52, 179

ICC/International Criminal Court 88
IMF/International Monetary Fund 30, 32, 33
IMT/International Monsanto Tribunal 88, 93–4, 113–14n1
IPL/Intellectual Property Laws 89
International Court of Justice 33, 142, 147, 164n34
International Red Cross 144

Monsanto 3, 4, 6, 85, 88–94, 97, 98, 105, 106, 113–14n1, 115n14, 117n42, 176, 181

ODA/Overseas Development Assistance 113OECD/Organisation for Economic Co-operation and Development 21, 31, 32
OPEC/Organisation of the Petroleum Exporting Countries 25–6, 44

Paris Agreement 25, 169
Plant Patent Act 105

SOCO 33–6
Syngenta 89, 95, 105, 113–14n1

Treaty of Friendship and Partnership Italy/Libya (2008) 156
TRIPS/WTO Agreement on Trade-Related Aspects of Intellectual Property 106

UN/United Nations 133, 142, 160; *UN Convention on the Rights of the Child* 124; *UN Development Plan* 113–14n1, 127; *UNESCO* 33, 131, 163; *UN General Assembly Economic and Social Council* 139; *UN Goldstone Mission Report* 146; *UN Human Rights Committee/Universal Human Rights* 142, 160; *UNHCR* 121–3, 132, 148, 149, 151–2, 160, 162n4–5, 162n7–9, 163n24, 165n56; *UN Refugee Convention and Protocol* 121–3, 124, 146–7, 149, 159, 160, 162n8, 166n62; *UN Food and Agriculture Organisation (FAO)* 85–7, 90, 100–1, 104, 106, 114n4, 117n55

WHO/World Health Organisation 32, 63
World Bank 33
WTO/World Trade Organisation 89, 101, 104, 105, 109, 118n65